DATE DUE

BOOKS MAY BE RECALLED
AFTER 10 DAYS

Printed
in USA

D0088943

Energy-Efficient Electric Motors

ELECTRICAL ENGINEERING AND ELECTRONICS

A Series of Reference Books and Textbooks

Editors

Marlin O. Thurston
Department of Electrical
Engineering
The Ohio State University
Columbus, Ohio

William Middendorf
Department of Electrical
and Computer Engineering
University of Cincinnati
Cincinnati, Ohio

1. Rational Fault Analysis, *edited by Richard Saeks and S. R. Liberty*

2. Nonparametric Methods in Communications, *edited by P. Papantoni-Kazakos and Dimitri Kazakos*

3. Interactive Pattern Recognition, *Yi-tzuu Chien*

4. Solid-State Electronics, *Lawrence E. Murr*

5. Electronic, Magnetic, and Thermal Properties of Solid Materials, *Klaus Schröder*

6. Magnetic-Bubble Memory Technology, *Hsu Chang*

7. Transformer and Inductor Design Handbook, *Colonel Wm. T. McLyman*

8. Electromagnetics: Classical and Modern Theory and Applications, *Samuel Seely and Alexander D. Poularikas*

9. One-Dimensional Digital Signal Processing, *Chi-Tsong Chen*

10. Interconnected Dynamical Systems, *Raymond A. DeCarlo and Richard Saeks*

11. Modern Digital Control Systems, *Raymond G. Jacquot*

12. Hybrid Circuit Design and Manufacture, *Roydn D. Jones*

13. Magnetic Core Selection for Transformers and Inductors: A User's Guide to Practice and Specification, *Colonel Wm. T. McLyman*

14. Static and Rotating Electromagnetic Devices, *Richard H. Engelmann*

15. Energy-Efficient Electric Motors: Selection and Application, *John C. Andreas*

Other Volumes in Preparation

Energy-Efficient Electric Motors

SELECTION AND APPLICATION

JOHN C. ANDREAS

Electric Motor Consultant
Retired Vice-President of Product Development
Electric Motor Division, Gould Inc.

MARCEL DEKKER, INC.   New York and Basel

Library of Congress Cataloging in Publication Data

Andreas, John C., [date]
  Energy-efficient electric motors.

  (Electrical engineering and electronics; 15)
  Bibliography: p.
  Includes index.
  1. Electric motors. I. Title. II. Series.
TK2511.A55   621.46′2   82-2354
ISBN 0-8247-1786-4   AACR2

MARCEL DEKKER, INC.
270 Madison Avenue, New York, New York 10016

Current printing (last digit):
10 9 8 7 6 5 4 3 2 1

PRINTED IN THE UNITED STATES OF AMERICA

To my wife
Ruth E. Andreas
for her faith and encouragement

Preface

The number of electric motors in the 1- to 125-hp range was approximately 70 million in 1977 and is increasing 6 percent per year according to a recent study by the U.S. Department of Energy. This study also noted that 53 to 58 percent of the electric energy generated is consumed by electric-motor-driven systems. This presents us with an opportunity to save considerable energy by wisely selecting motors and the devices they drive. However, it should be recognized that the electric motor is a device for converting electrical energy to rotating mechanical energy. The only power consumed by the electric motor is the electrical and mechanical energy losses within the motor, and the balance of the electrical energy is transferred as mechanical energy to some driven device such as a pump, fan, or conveyor. Since the motor losses are 5 to 25 percent of the input power, it is important to consider the complete system, including the electric motor, when determining system efficiency and potential energy conservation.

Modern motors are precisely designed, taking advantage of computer-derived optimum designs, high-quality materials, and improved manufacturing technology. Hence, for many years the trend was toward smaller and lighter motors in order to lower cost, and no significant attention was given to efficiency and the power factor beyond the levels required to achieve allowable temperatures.

With the increasing cost of electric power, in 1975 motor manufacturers began addressing the problem of improving electric motor efficiencies to levels that would represent significant savings in energy.

Coincident with the trend toward smaller motors, many users

and original equipment manufacturers have been choosing to purchase the lowest-first-cost motor without considering the power factor and efficiency. Similarly, many textbooks and handbooks on electric motors discuss in great detail the design and performance characteristics of electric motors and the characteristics of various types of motor loads and how to match the motor to the load requirements. However, efficiency, the power factor, energy costs, and life-cycle costing have not been considered as major factors in the selection of an electric motor in most applications.

In many cases, electric motors have been selected and applied by engineers or other personnel who have a limited knowledge of electric motors, particularly a lack of understanding of the power factor, efficiency, and associated energy economies.

Today, with the high cost of electrical energy and the continuing trend toward higher costs, electric motors should be applied and selected on a life-cycle cost basis, including such factors as first cost, energy efficiency, the duty cycle, operating time, and energy costs.

My goal in this book is to provide guidelines for selecting and applying electric motors on an energy conservation and life-cycle cost basis. Particular emphasis is given to both single-phase and three-phase motors in the 1- to 125-hp range since this is the range that offers the maximum opportunities for energy savings. It is my intention to present these guidelines in a format that can be understood and effectively used by all personnel responsible for the application, selection, and procurement of electric motors, motor controls, and motor-driven products.

John C. Andreas

Contents

Energy-Efficient Electric Motors

1
Induction Motor Characteristics

1.1 THREE-PHASE INDUCTION MOTORS

In the integral horsepower sizes, i.e., above 1 hp, three-phase induction motors of various types drive more industrial equipment than any other means. The most common three-phase (polyphase) induction motors fall within the following major types:

NEMA* design B: Normal torques, normal slip, normal locked amps
NEMA design A: High torques, low slip, high locked amps
NEMA design C: High torques, normal slip, normal locked amps
NEMA design D: High locked-rotor torque, high slip
Wound rotor: Characteristics depend on external resistance
Multispeed: Characteristics depend on design—variable torque, constant torque, constant horsepower

There are many specially designed electric motors with unique characteristics to meet specific needs. However, the majority of needs can be met with the preceding motors.

NEMA Design B Motors

The NEMA design B motor is the basic integral horsepower motor. It is a three-phase motor designed with normal torque and normal starting current and generally has a slip at the rated load of less than 4 percent. Thus the motor speed in revolutions per minute is 96 percent or more

*National Electrical Manufacturers Association.

1

of the synchronous speed for the motor. For example, a four-pole motor operating on a 60-Hz line frequency has a synchronous speed of 1800 rpm or a full-load speed of

$$1800 - (1800 \times \text{slip}) = 1800 - (1800 \times 0.4)$$
$$= 1800 - 72$$
$$= 1728 \text{ rpm}$$

or

$$1800 \times 0.96 = 1728 \text{ rpm}$$

In general, most three-phase motors in the 1- to 125-hp range have a slip at the rated load of approximately 3 percent, or in the case of four-pole motors, a full-load speed of 1745 rpm. Figure 1.1 shows the typical construction for a totally enclosed, fan-cooled NEMA design B motor with a die-cast aluminum single-cage rotor.

Figure 1.2 shows the typical speed-torque curve for the NEMA design B motor. This type of motor has moderate starting torque, a

Fig. 1.1 NEMA design B totally enclosed, fan-cooled polyphase induction motor. (Courtesy of Gould Inc., Electric Motor Division, St. Louis.)

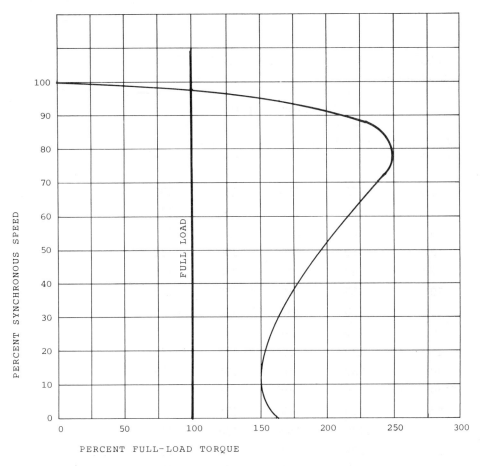

Fig. 1.2 NEMA design B motor speed-torque curve.

pull-up torque exceeding the full-load torque, and a breakdown torque (or maximum torque) several times the full-load torque. Thus it can provide starting and smooth acceleration for most loads and in addition can sustain temporary peak loads without stalling. The NEMA performance standards for design B motors are shown in Table 1.1, Table 1.2, and Table 1.3. There are no established standards for the efficiency or the power factor of NEMA design B motors. However, NEMA has established standards for testing and labeling motors; these standards are discussed in detail in Chap. 2.

Table 1.1   Locked-Rotor Torque of NEMA Design A
and B Motors[a,b]

hp	Synchronous speed, 60 Hz			
	3600 rpm	1800 rpm	1200 rpm	900 rpm
1	—	275	170	135
1.5	175	250	165	130
2	170	235	160	130
3	160	215	155	130
5	150	185	150	130
7.5	140	175	150	125
10	135	165	150	120
15	130	160	140	125
20	130	150	135	125
25	130	150	135	125
30	130	150	135	125
40	125	140	135	125
50	120	140	135	125
60	120	140	135	125
75	105	140	135	125
100	105	125	125	125
125	100	110	125	120
150	100	110	120	120
200	100	100	120	120
250	70	80	100	100

[a]Single-speed, polyphase, squirrel-cage, integral horsepower
motors with continuous ratings (percent of full-load torque).
[b]For other speeds and ratings, see NEMA Standard MG1-12.37.
Source: Reprinted by permission from NEMA Standards Publication No. MG1-1978, *Motors and Generators,* copyright 1978
by the National Electrical Manufacturers Association.

Table 1.2 Breakdown Torque of NEMA Design
A and B Motors[a,b]

hp	Synchronous speed, 60 Hz			
	3600 rpm	1800 rpm	1200 rpm	900 rpm
1	—	300	265	215
1.5	250	280	250	210
2	240	270	240	210
3	230	250	230	205
5	215	225	215	205
7.5	200	215	205	200
10	200	200	200	200
15	200	200	200	200
20	200	200	200	200
25	200	200	200	200
30	200	200	200	200
40	200	200	200	200
50	200	200	200	200
60	200	200	200	200
75	200	200	200	200
100	200	200	200	200
125	200	200	200	200
150	200	200	200	200
200	200	200	200	200
250	175	175	175	175

[a]Single-speed, polyphase, squirrel-cage, integral horsepower
motors with continuous ratings (percent of full-load torque).
[b]For other speeds and ratings, see NEMA Standard MG1-12.38.

Source: Reprinted by permission from NEMA Standards Publication No. MG1-1978, *Motors and Generators,* copyright 1978
by the National Electrical Manufacturers Association.

Table 1.3 Locked-Rotor Current of NEMA
Design B, C, and D Motors[a,b]

hp	Locked-rotor current (A)[c]	NEMA design letter	Code letter
1	30	B, D	N
1.5	40	B, D	M
2	50	B, D	L
3	64	B, C, D	K
5	92	B, C, D	J
7.5	127	B, C, D	H
10	162	B, C, D	H
15	232	B, C, D	G
20	290	B, C, D	G
25	365	B, C, D	G
30	435	B, C, D	G
40	580	B, C, D	G
50	725	B, C, D	G
60	870	B, C, D	G
75	1085	B, C, D	G
100	1450	B, C, D	G
125	1815	B, C, D	G
150	2170	B, C, D	G
200	2900	B, C	G
250	3650	B	G

[a]Three-phase, 60-Hz, integral horsepower, squirrel-cage induction motors rated at 230 V.
[b]For other horsepower ratings, see NEMA Standard MG1-12.34.
[c]The locked-rotor current for motors designed for voltages other than 230 V shall be inversely proportional to the voltage.

Source: Reprinted by permission from NEMA Standards publication No. MG1-1978, *Motors and Generators,* copyright 1978 by the National Electrical Manufacturers Association.

NEMA Design A Motors

The NEMA design A motor is a polyphase, squirrel-cage induction motor designed with torques and locked-rotor currents that exceed the corresponding values for NEMA design B motors. The criterion for classification as a design A motor is that the value of the locked-rotor current be in excess of the value for NEMA design B motors. The NEMA design A motor is usually applied to special applications that cannot be served by NEMA design B motors, and most often these applications require motors with higher than normal breakdown torques to meet the requirements of high transient or short-duration loads. The NEMA design A motor is also applied to loads requiring extremely low slip, on the order of 1 percent or less.

NEMA Design C Motors

The NEMA design C motor is a squirrel-cage induction motor that develops high locked-rotor torques for hard-to-start applications. Figure 1.3 shows the construction of a dripproof NEMA design C motor with a double-cage, die-cast aluminum rotor. Figure 1.4 shows the typical speed-torque curve for the NEMA design C motor. These motors have a slip at the rated load of less than 5 percent. The NEMA performance standards for NEMA design C motors are shown in Table 1.4, Table 1.5, and Table 1.3.

NEMA Design D Motors

The NEMA design D motor combines high locked-rotor torque with high full-load slip. Two standard designs are generally offered, one with full-load slip of 5 to 8 percent and the other with full-load slip of 8 to 13 percent. The locked-rotor torque for both types is generally 275 to 300 percent of full-load torque; however, for special applications the locked-rotor torque can be higher. Figure 1.5 shows the typical speed-torque curves for NEMA design D motors. These motors are recommended for cyclical loads such as those found in punch presses that have stored energy systems in the form of flywheels to average the motor load and are excellent for loads of short duration with frequent starts and stops. The proper application of this type of motor requires detailed information about the system inertia, duty cycle, and operating load as well as the motor characteristics. With this information, the motors are selected and applied on the basis of their thermal capacity.

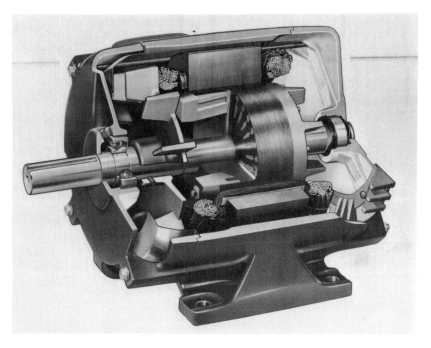

Fig. 1.3 NEMA design C dripproof polyphase induction motor. (Courtesy of Gould Inc., Electric Motor Division, St. Louis.)

Wound Rotor Induction Motors

The wound rotor induction motor is an induction motor in which the secondary (or rotating) winding is an insulated polyphase winding similar to the stator winding. The rotor winding generally terminates at collector rings on the rotor, and stationary brushes are in contact with each collector ring to provide access to the rotor circuit. There are a number of systems available to control the secondary resistance of the motor and hence the motor's characteristics. The use and application of wound rotor induction motors have been mostly limited to hoist and crane applications and special speed control applications. Typical wound rotor motor speed-torque curves for various values of resistance inserted in the rotor circuit are shown in Fig. 1.6. As the value of resistance is increased, the characteristic of the speed-torque curve progresses from curve 1 with no external resistance to curve 4 with high external resistance. With appropriate control equipment, the characteristics of the motor can be changed by changing this value of external rotor resistance. Solid-state inverter systems have been developed which, when connected in the rotor circuit instead of resistors, return the slip loss of the motor to the power line. This system substantially improves the efficiency of the wound rotor motor when used in variable-speed applications.

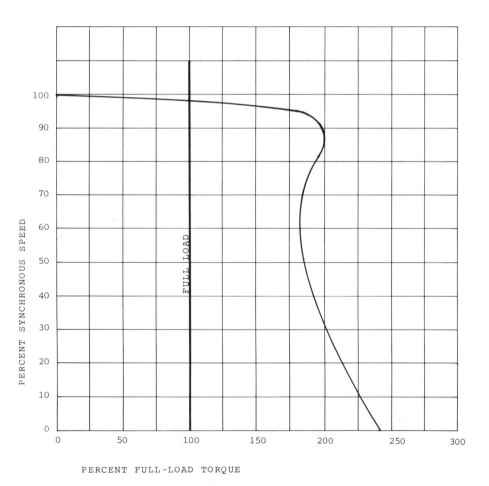

Fig. 1.4   NEMA design C motor speed-torque curve.

Table 1.4 Locked-Rotor Torque of NEMA Design C Motors[a]

hp	Synchronous speed, 60 Hz		
	1800 rpm	1200 rpm	900 rpm
3	—	250	225
5	250	250	225
7.5	250	225	200
10	250	225	200
15	225	200	200
20–200 Inclusive	200	200	200

[a]Single-speed, polyphase, squirrel-cage, integral horsepower motors with continuous ratings (percent of full-load torque).

Source: Reprinted by permission from NEMA Standards Publication No. MG1-1978, *Motors and Generators*, copyright 1978 by the National Electrical Manufacturers Association.

Table 1.5 Breakdown Torque of NEMA Design C Motors[a]

hp	Synchronous speed, 60 Hz		
	1800 rpm	1200 rpm	900 rpm
3	—	225	200
5	200	200	200
7.5–200 Inclusive	190	190	190

[a]Single-speed, polyphase, squirrel-cage, integral horsepower motors with continuous ratings (percent of full-load torque).

Source: Reprinted by permission from NEMA Standards Publication No. MG1-1978, *Motors and Generators*, copyright 1978 by the National Electrical Manufacturers Association.

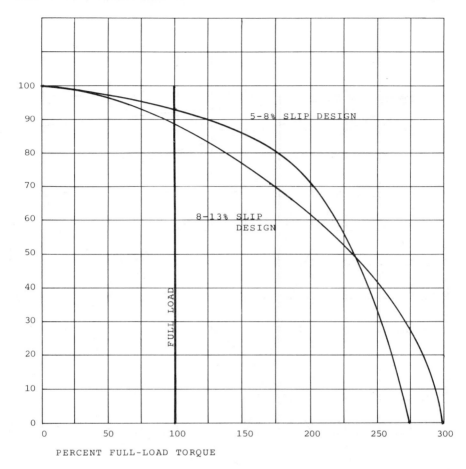

Fig. 1.5   NEMA design D motor speed-torque curves: 5 to 8 percent and 8 to 13 percent slip.

Multispeed Motors

Multispeed motors, i.e., motors that will operate at more than one speed, with characteristics similar to those of the NEMA-type motors are also available. The multispeed induction motors usually have one or two primary windings. In one-winding motors, the ratio of the two speeds must be 2 to 1; for example, possible speed combinations are 3600/1800, 1800/900, and 1200/600 rpm. In two-winding motors, the ratio of the speeds can be any combination within certain design limits,

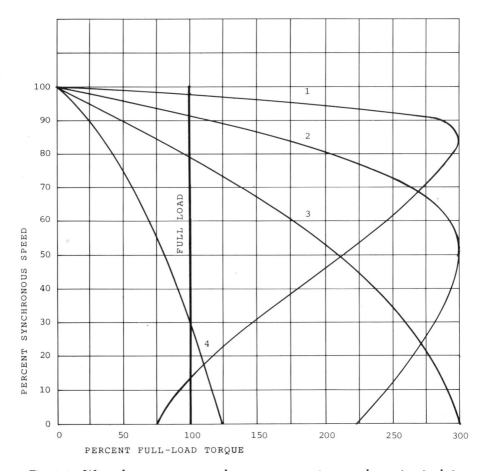

Fig. 1.6 Wound rotor motor speed-torque curves: 1, rotor short-circuited; 2 to 4, increasing values of external resistance.

depending on the number of winding slots in the stator. The most popular combinations are 1800/1200, 1800/900, and 1800/600 rpm. In addition, two-winding motors can be wound to provide two speeds on each winding; this makes it possible for the motor to operate at four speeds, for example, 3600/1800 rpm on one winding and 1200/600 rpm on the other winding.

Multispeed motors are available with the following torque characteristics.

Variable Torque. The variable-torque multispeed motor has a torque output which varies directly with the speed, and hence the horsepower output varies with the square of the speed. This motor is commonly used with fans, blowers, and centrifugal pumps to control the output of the driven device. Figure 1.7 shows typical speed-torque curves for this type of motor. Superimposed on the motor speed-torque curve is the speed-torque curve for a typical fan where the input horsepower to the fan varies as the cube of the fan speed. Another popular

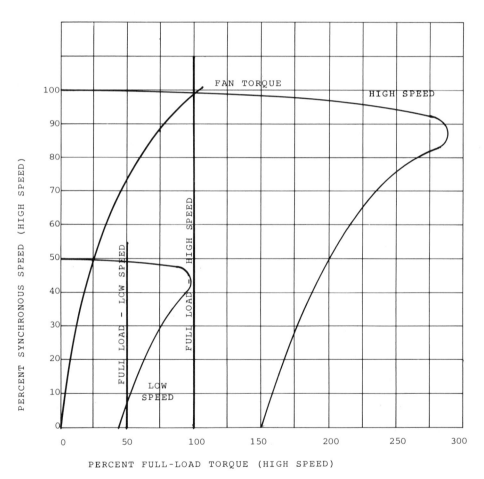

Fig. 1.7 Speed-torque curves for a variable-torque, one-winding, two-speed motor.

drive for fans is a two-winding two-speed motor, such as 1800 rpm at high speed and 1200 rpm at low speed. Figure 1.8 shows the typical motor speed-torque curve for the two-winding variable-torque motor with a fan speed-torque curve superimposed.

   Constant Torque.   The constant-torque multispeed motor has a torque output which is the same at all speeds, and hence the horsepower output varies directly with the speed. This motor can be used with friction-type loads such as those found on conveyors to control the conveyor speed. Figure 1.9 shows typical speed-torque curves.

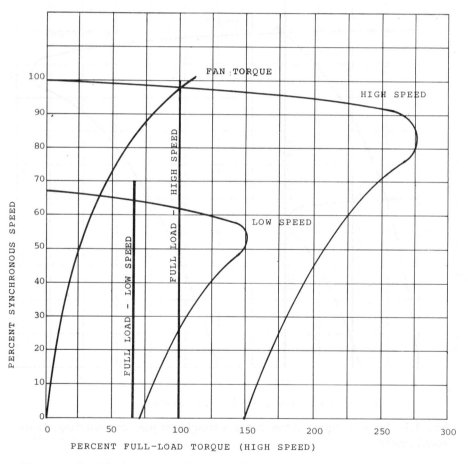

Fig. 1.8   Speed-torque curves for a multispeed variable-torque motor with two windings, two speeds, and a four-pole to six-pole ratio.

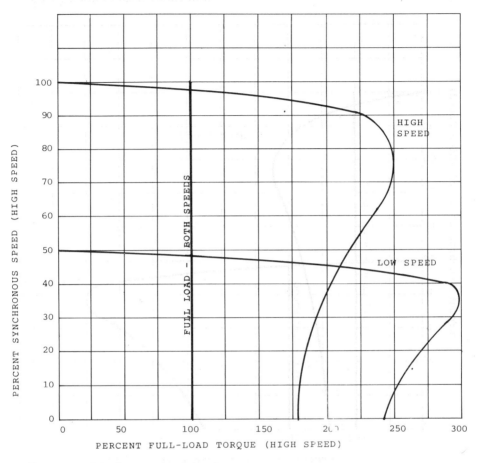

Fig. 1.9 Speed-torque curves for a constant-torque, one-winding, two-speed motor.

Constant Horsepower. The constant-horsepower multispeed motor has the same horsepower output at all speeds. This type of motor is used for machine tool applications that require higher torques at lower speeds. Figure 1.10 shows typical speed-torque curves.

1.2 SINGLE-PHASE INDUCTION MOTORS

There are many types of single-phase electric motors. In this section the discussion will be limited to those types most common to integral horsepower motor ratings of 1 hp and higher.

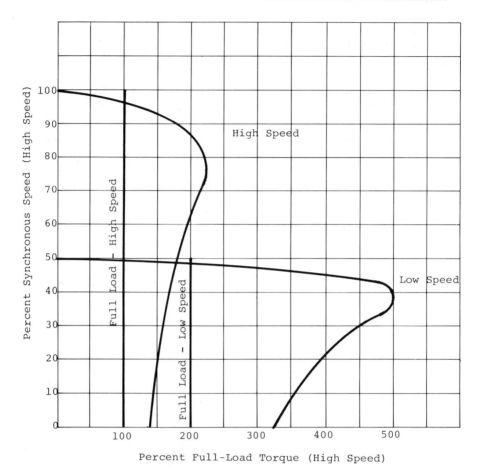

Fig. 1.10 Speed-torque curves for a constant-horsepower one-winding, two-speed motor.

In industrial applications, three-phase induction motors should be used wherever possible. In general, three-phase electric motors have higher efficiency and power factors and are more reliable since they do not have starting switches or capacitors.

In those instances when three-phase electric motors are not available or cannot be used because of the power supply, the following types of single-phase motors are recommended for industrial and commercial applications: (1) capacitor start motor, (2) two-value capacitor motor, and (3) permanent split capacitor motor.

A brief comparison of single-phase and three-phase induction motor characteristics will provide a better understanding of how single-phase motors perform:

1. Three-phase motors have locked torque because there is a revolving field in the air gap at standstill. A single-phase motor has no revolving field at standstill and therefore develops no locked-rotor torque. An auxiliary winding is necessary to produce the rotating field required for starting. In an integral horsepower single-phase motor, this is part of an RLC network.

2. The rotor current and rotor losses are insignificant at no load in a three-phase motor. Single-phase motors have appreciable rotor current and rotor losses at no load.

3. For a given breakdown torque, the single-phase motor requires considerably more flux and more active material than the equivalent three-phase motor.

4. A comparison of the losses between single-phase and three-phase motors is shown in Fig. 1.11. Note the significantly higher losses in the single-phase motor.

The general characteristics of these types of single-phase induction motors are as follows.

Capacitor Start Motors

A capacitor start motor is a single-phase induction motor with a main winding arranged for direct connection to the power source and an auxiliary winding connected in series with a capacitor and starting switch for disconnecting the auxiliary winding from the power source after starting. Figure 1.12 is a schematic diagram of a capacitor start motor. The type of starting switch most commonly used is a centrifugally actuated switch built into the motor. Figure 1.13 illustrates an industrial-quality dripproof single-phase capacitor start motor; note the centrifugally actuated switch mechanism.

However, other types of devices such as current-sensitive and voltage-sensitive relays are also used as starting switches. More recently, solid-state switches have been developed and used to a limited extent. The solid-state switch will be the switch of the future as it is refined and costs are reduced.

All the switches are set to stay closed and maintained the auxili-

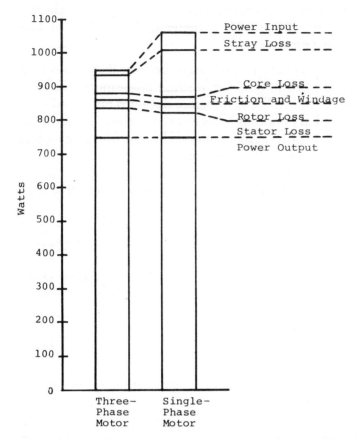

Fig. 1.11   Motor loss comparison of three-phase and single-phase motors.

ary winding circuit in operation until the motor starts and accelerates to approximately 80 percent of full-load speed. At that speed, the switch opens, disconnecting the auxiliary winding circuit from the power source.

The motor then runs on the main winding as an induction motor. The typical speed-torque characteristics for a capacitor start motor are shown in Fig. 1.14. Note the change in motor torques at the transition point when the starting switch operates.

The typical performance data for integral horsepower, 1800-rpm, capacitor start, induction run motors are shown in Table 1.6. There will be a substantially wider variation in the values of locked-rotor torque, breakdown torque, and pull-up torque for these single-phase motors than for comparable three-phase motors, and the same varia-

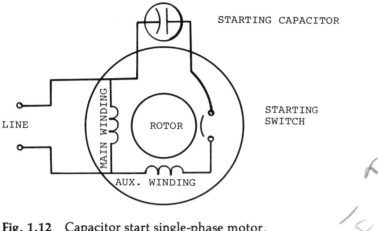

Fig. 1.12   Capacitor start single-phase motor.

Fig. 1.13   Capacitor start single-phase motor. (Courtesy of Gould Inc., Electric Motor Division, St. Louis.)

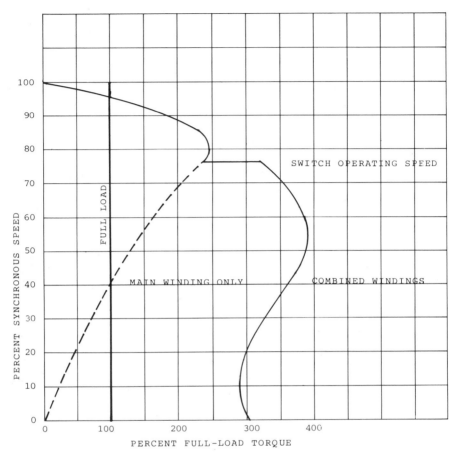

Fig. 1.14   Speed-torque curve for a capacitor start motor.

tion also exists for efficiency and the power factor (PF). Note that pull-up torque is a factor in single-phase motors to ensure starting with high inertia or hard-to-start loads. Therefore, it is important to know the characteristics of the specific capacitor start motor to make certain it is suitable for the application.

Two-Value Capacitor Motors

A two-value capacitor motor is a capacitor motor with different values of capacitance for starting and running. Very often this type of motor is referred to as a capacitor start, capacitor run motor.

The change in the value of capacitance from starting to running conditions is automatic by means of a starting switch the same as is used for the capacitor start motors. Two capacitors are provided, a high value of capacitance for starting conditions and a lower value for running conditions. The starting capacitor is usually an electrolytic type which provides high capacitance per unit volume. The running capacitor is usually a paper-spaced, oil-filled unit rated for continuous operation. Figure 1.15 shows one method of mounting both capacitors on the motor.

The schematic diagram for a two-value capacitor motor is shown in Fig. 1.16. As shown, at starting, both the starting and running capacitors are connected in series with the auxiliary winding. When the starting switch opens, it disconnects the starting capacitor from the auxiliary winding circuit but leaves the running capacitor in series with the auxiliary winding connected to the power source. Thus both the main and auxiliary windings are energized when the motor is running and contribute to the motor output. A typical speed-torque curve for a two-value capacitor motor is shown in Fig. 1.17.

For a given capacitor start motor, the effect of adding a running capacitor in the auxiliary winding circuit is as follows:

Increased breakdown torque: 5 to 30 percent
Increased lock-rotor torque: 5 to 10 percent
Improved full-load efficiency: 2 to 7 points
Improved full-load power factor: 10 to 20 points

Table 1.6 Typical Performance of Capacitor Start Motors[a]

| hp | Full-load performance | | | | | Torques | | |
	rpm	A	Eff.	PF	Torque	Locked	Break-down	Pull up
1	1725	7.5	71	70	3.0	9.9	7.5	7.6
2	1750	12.5	72	72	6.0	17.5	14.7	11.5
3	1750	17.0	74	79	9.0	23.0	21.0	18.5
5	1745	27.3	78	77	15.0	46.0	32.0	35.0

[a]Four-pole, 230-V, single-phase motors.
Source: Courtesy of Gould Inc., Electric Motor Division, St. Louis, Mo.

Fig. 1.15 Two-value capacitor single-phase motor. (Courtesy of Gould Inc., Electric Motor Division, St. Louis.)

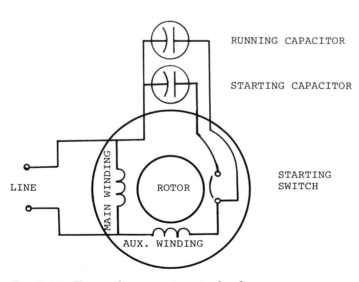

Fig. 1.16 Two-value capacitor single-phase motor.

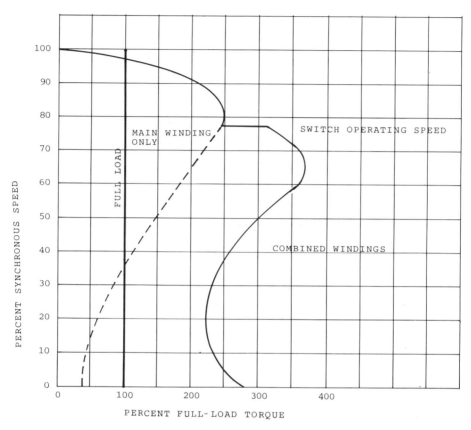

Fig. 1.17 Speed-torque curve for a two-value capacitor motor.

Reduced full-load running current
Reduced magnetic noise
Cooler running

The addition of a running capacitor to a single-phase motor with properly designed windings permits the running performance to approach the performance of a three-phase motor. The typical performance of integral horsepower two-value capacitor motors is shown in Table 1.7. Comparison of this performance with the performance shown in Table 1.6 for capacitor start motors shows the improvement in both efficiency and the power factor.

The optimum performance that can be achieved in a two-value

Table 1.7 Typical Performance of Two-Value Capacitor Motors[a]

hp	Full-load performance				Torques			
	rpm	A	Eff.	PF	Torque	Locked	Break-down	Pull up
3	1760	14.0	78	90	9.0	25	23	22
5	1760	25.0	82	80	15.0	46	35	32
7.5	1750	32.0	86	88	22.5	45	56	45
10	1750	38.0	86	96	30.0	56	72	56

[a]Four-pole, 230-V, single-phase motors.
Source: Courtesy of Gould Inc., Electric Motor Division, St. Louis.

capacitor single-phase motor is a function of the economic factors as well as the technical considerations in the design of the motor. To illustrate this, Table 1.8 shows the performance of a single-phase motor with the design optimized for various values of running capacitance. The base for the performance comparison is a capacitor start, induc-

Table 1.8 Performance Comparison of Capacitor Start and Two-Value Capacitor Motors

	Type of motor				
	Capacitor start	Two-value capacitor			
Running capacitor, MFD	0	7.5	15	30	65
Full-load efficiency	70	78	79	81	83
Full-load PF	79	94	97	99[a]	99[a]
Input watts reduction, %	0	10.1	11.5	13.3	15
Cost, %	100	130	140	151	196
Approximate payback period	—	1.3	1.6	1.8	2.9

[a]Leading power factor.

tion-run motor with no running capacitor. Table 1.8 shows that performance improves with increasing values of running capacitance and that the motor costs increase as the value of running capacitance is increased. The payback period in years was calculated on the basis of 4000 hr/yr of operation and an electric power cost of 6¢/kWh. Note that the major improvement in motor performance is made in the initial change from a capacitor start to a two-value capacitor motor with a relatively low value of running capacitance. This initial design change also shows the shortest payback period.

The determination of the optimum two-value capacitor motor for a specific application requires a comparison of the motor costs and the energy consumptions of all such available motors. It is recommended that this comparison be made by a life-cycle cost method or the net present worth method (outlined in Chap. 7).

The efficiency improvement and energy savings of a specific product line of pool pump motors when the design was changed from capacitor start motors to two-value capacitor motors are illustrated by Figs. 1.18 and 1.19. Based on the same operating criterion used above, i.e., 4000-hr/yr operation at power costs of 6¢/kWh, the payback period for these motors was 8 to 20 months.

Permanent Split Capacitor Motors

The permanent split capacitor motor, a single-phase induction motor, is defined as a capacitor motor with the same value of capacitance used for both starting and running operations. This type of motor is also referred to as a single-value capacitor motor. The application of this type of single-phase motor is normally limited to the direct drive of such loads as those of fans, blowers, or pumps that do not require normal or high starting torques. Consequently, the major application of the permanent split capacitor motor has been to direct-driven fans and blowers. These motors are not suitable for belt-driven applications and are generally limited to the lower horsepower ratings.

The schematic diagram for a permanent split capacitor motor is shown in Fig. 1.20. Note the absence of any starting switch. This type of motor is essentially the same as a two-value capacitor motor operating on the running connection and will have approximately the same torque characteristics. Since only the running capacitor (which is of relative low value) is connected in series with the auxiliary winding on

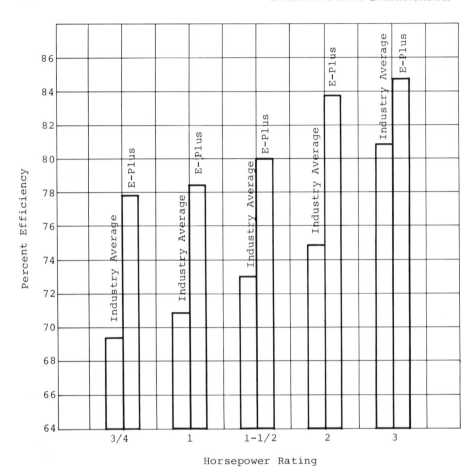

Fig. 1.18 Efficiency comparison of the Gould E-Plus energy-efficient pool pump motors versus the industry average. (Courtesy of Gould Inc., Electric Motors Division, St. Louis.)

starting, the starting torque is greatly reduced. The starting torque is only 20 to 30 percent of full-load torque. A typical speed-torque curve for a permanent split capacitor motor is shown in Fig. 1.21. The running performance of this type of motor in terms of efficiency and power factor is the same as a two-value capacitor motor. However, because of its low starting torque, its successful application requires close coordination between the motor manufacturer and the manufacturer of the driven equipment.

A special version of the capacitor motor is used for multiple-speed fan drives. This type of capacitor motor usually has a tapped main winding and a high-resistance rotor. The high-resistance rotor is used to improve stable speed operation and to increase the starting torque. There are a number of versions and methods of winding motors. The most common design is the two-speed motor which has three windings: the main winding, the intermediate winding, and the auxiliary winding. For 230-V power service, a common connection of the windings is called the T connection. Schematic diagrams for two-speed T-connected motors are shown in Figs. 1.22 and 1.23. For high-speed operation, the intermediate winding is not connected in the circuit as

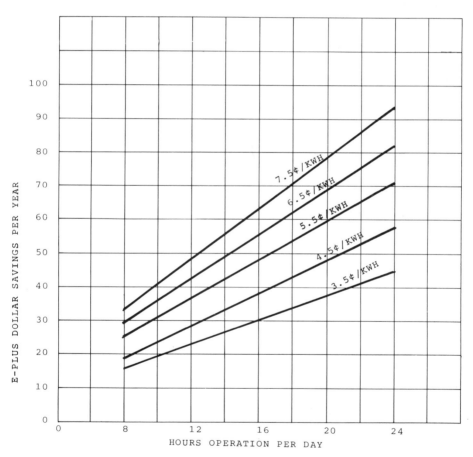

Fig. 1.19 Typical annual savings of a 1-hp motor (assume 365 days per year · operation). (Courtesy of Gould Inc., Electric Motor Division, St. Louis.)

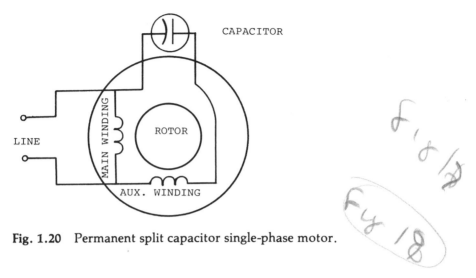

Fig. 1.20   Permanent split capacitor single-phase motor.

shown in Fig. 1.23, and line voltage is applied to the main winding and to the auxiliary winding and capacitor in series. For low-speed operation, the intermediate winding is connected in series with the main winding and with the auxiliary circuit as shown in Fig. 1.23. This connection reduces the voltage applied across both the main winding and the auxiliary circuit, thus reducing the torque the motor will develop and hence the motor speed to match the load requirements. The amount of speed reduction is a function of the turns ratio between the main and intermediate windings and the speed-torque characteristics of the driven load. It should be recognized that with this type of motor the speed change is obtained by letting the motor speed slip down to the required low speed; it is not a multispeed motor with more than one synchronous speed.

An example of the speed-torque curves for a tapped winding capacitor motor is shown in Fig. 1.24. The load curve of a typical fan load is superimposed on the motor speed-torque curves to show the speed reduction obtained on the low-speed connection.

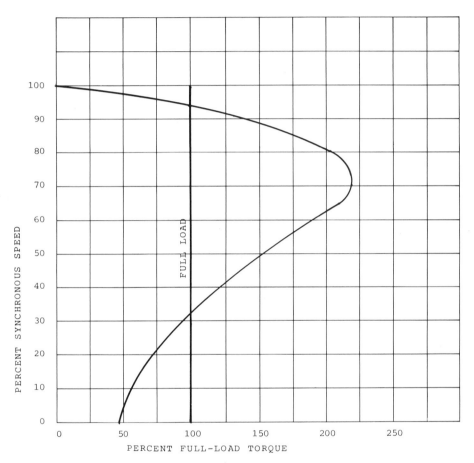

Fig. 1.21 Speed-torque curve for a permanent split capacitor motor.

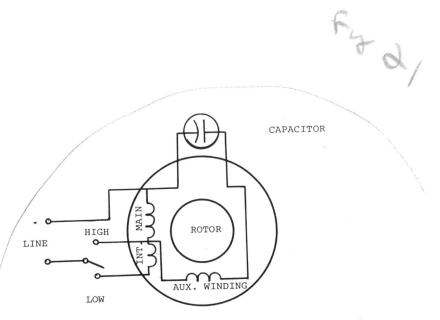

Fig. 1.22 Permanent split capacitor single-phase motor with a T-type connection and two-speed operation.

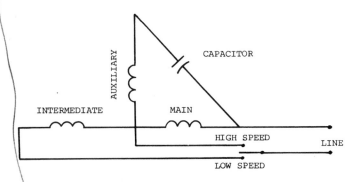

Fig. 1.23 Permanent split capacitor single-phase motor with a T-type connection and a winding arrangement.

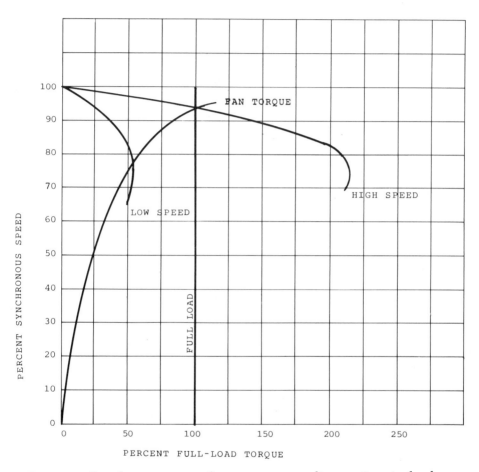

Fig. 1.24 Speed-torque curves for a permanent split capacitor single-phase motor with a tapped winding.

2
Energy-Efficient Motors

2.1 STANDARD MOTOR EFFICIENCY

During the period from 1960 to 1975, electric motors, particularly those in the 1- to 250-hp range, were designed for minimum first cost. The amount of active material, i.e., lamination steel, copper or aluminum or magnet wire, and rotor aluminum, was selected as the minimum levels required to meet the performance requirements of the motor. Efficiency was maintained at levels high enough to meet the temperature rise requirements of the particular motor. As a consequence, depending on the type of enclosure and ventilation system, a wide range in efficiencies exists for standard NEMA design B polyphase motors. Table 2.1 is an indication of the range of the nominal electric motor efficiencies at rated horsepower. These data are also presented in Fig. 2.1. The data are based on information published by the major electric motor manufacturers. However, the meaning or interpretation of data published prior to the NEMA adoption of the definition of nominal efficiency is not always clear. In 1977, NEMA recommended a procedure for marking the three-phase motors with a *NEMA nominal efficiency*. This efficiency represents the average efficiency for a large population of motors of the same design. In addition, a minimum efficiency was established for each level of nominal efficiency.

The minimum efficiency is the lowest level of efficiency to be expected when a motor is marked with the nominal efficiency in accordance with the NEMA standard. This method of identifying the motor

Table 2.1 Full-Load Efficiencies of NEMA Design B Standard Three-Phase Induction Motors

hp	Nominal efficiency range	Average nominal efficiency
1	68–78	73
1.5	68–80	75
2	72–81	77
3	74–83	80
5	78–85	82
7.5	80–87	84
10	81–88	85
15	83–89	86
20	84–89	87.5
25	85–90	88
30	86–90.5	88.5
40	87–91.5	89.5
50	88–92	90
60	88.5–92	90.5
75	89.5–92.5	91
100	90–93	91.5
125	90.5–93	92
150	91–93.5	92.5
200	91.5–94	93
250	91.5–94.5	93.5

efficiency takes into account variations in materials, manufacturing processes, and test results in motor-to-motor efficiency variations for a given motor design. The nominal efficiency represents a value which should be used to compute the energy consumption of a motor or group of motors. Table 2.1 shows a wide range in efficiency for individual motors and consequently a range in the electric motor losses

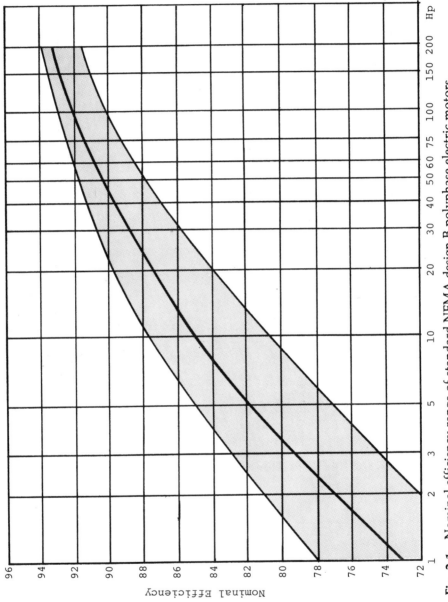

Fig. 2.1 Nominal efficiency range of standard NEMA design B polyphase electric motors.

and electric power input. For example, a standard 10 hp electric motor may have an efficiency range of 81 to 88 percent. *At 81% efficiency,*

$$\text{Input electric power} = \frac{10 \times 746}{0.81} = 9210 \text{ W}$$

$$\text{Motor losses} = 9210 - 7460 = 1750 \text{ W}$$

At 88% efficiency,

$$\text{Input electric power} = \frac{10 \times 746}{0.88} = 8477 \text{ W}$$

$$\text{Motor losses} = 8477 - 7460 = 1017 \text{ W}$$

Therefore, for the same output the input can range from 8477 to 9210 W or an increase in energy consumption and power costs of 8 percent to operate the less efficient motor.

2.2 WHY MORE EFFICIENT MOTORS?

The escalation of oil prices and the cost of electric power beginning in 1974 makes it increasingly expensive to use inefficient electric motors. Since 1972, electric power rates have increased at an average of 11.5 percent per year. The annual power cost to operate a 10-hp motor 4000 hr/yr has increased from $850 in 1972 to $1950 in 1980. Consequently, by 1974 electric motor manufacturers were looking for methods to improve three-phase induction motor efficiencies to values above those shown for standard NEMA design B motors in Table 2.1.

2.3 WHAT IS EFFICIENCY?

Electric motor efficiency is the measure of the ability of an electric motor to convert electrical energy to mechanical energy; i.e., kilowatts of electric power are supplied to the motor at its electrical terminals and the horsepower of mechanical energy is taken out of the motor at the rotating shaft. Therefore, the only power absorbed by the electric motor is the losses incurred in making the conversion from electrical to mechanical energy. Thus the motor efficiency can be expressed as

$$\text{Efficiency} = \frac{\text{mechanical energy out}}{\text{electrical energy in}} \times 100 \text{ percent}$$

but

Mechanical energy out = electrical energy in − motor losses

or

Electrical energy in = mechanical energy out + motor losses

Therefore, to reduce the electric power consumption for a given mechanical energy out, the motor losses must be reduced or the electric motor efficiency increased.

To accomplish this, it is necessary to understand the types of losses that occur in an electric motor. These losses consist of the following:

Power Losses

The power losses (I^2R in the motor windings) consist of two losses: the stator power losses I^2R and the rotor power losses I^2R. The stator power loss is a function of the current flowing in the stator winding and the stator winding resistance—hence the term I^2R loss:

$$\text{Stator current } I_1 = \frac{\text{input watts}}{\text{voltage} \times \sqrt{3}\,\text{PF}}$$

When improving the motor performance, it is important to recognize the interdependent relationship of the efficiency and the power factor. Rewrite the preceding equation and solve for the power factor:

$$\text{PF} = \frac{\text{output hp} \times 746}{\text{voltage} \times \sqrt{3}\,\text{efficiency} \times I_1}$$

Therefore, if the efficiency is increased, the power factor will tend to decrease. For the power factor to remain constant, the stator current I_1 must decrease in proportion to the increase in efficiency. To increase the power factor, the stator current must be decreased more than the efficiency is increased. From a design standpoint, this is difficult to accomplish and still maintain other performance requirements such as breakdown torque. However,

$$\text{Input watts} = \frac{\text{output hp} \times 746}{\text{efficiency}}$$

or

$$I_1 = \frac{\text{output hp} \times 746}{\text{voltage} \times \sqrt{3}\, PF \times \text{efficiency}}$$

Therefore, the stator losses are inversely proportional to the square of the efficiency and the power factor. In addition, the stator loss is a function of the stator winding resistance. For a given configuration, the winding resistance R is inversely proportional to the pounds of magnet wire or conductors in the stator winding. The more conductor material in the stator winding, the lower the losses.

The rotor power loss is generally expressed as the slip loss:

$$\text{Rotor loss} = \frac{(\text{hp output} \times 746 + FW)S}{1 - S}$$

$$S = \frac{N_s - N}{N} = \text{slip}$$

where

N = output speed, rpm
N_s = synchronous speed, rpm
FW = friction and windage loss

The rotor slip can be reduced by increasing the amount of conductor material in the rotor or increasing the total flux across the air gap into the rotor. The extent of these changes is limited by the minimum starting (or locked-rotor) torque required, the maximum locked-rotor current, and the minimum power factor required.

Magnetic Core Losses

Magnetic core losses consist of the eddy current and hysteresis losses, including the surface losses, in the magnetic structure of the motor. A number of factors influence these losses:

1. The flux density in the magnetic structure is a major factor in determining these magnetic losses. The core loss can be decreased by increasing the length of the magnetic structure and, as a consequence, decreasing the flux density in the core. This will decrease the magnetic loss per unit of weight, but since the total weight will increase, the improvement in losses will not be proportional to the unit loss reduction. The decrease in magnetic loading in the motor also decreases the magnetizing current and thus influences the power factor.
2. The magnetic core loss can also be reduced by using thinner lami-

nations in the magnetic structure. Typically, many standard motors use 24-gauge (0.025-thick) laminations. By using thinner laminations such as 26 gauge (0.0185 thick) or 29 gauge (0.014 thick), the magnetic core loss can be reduced. The reduction in the magnetic core loss by using thinner laminations ranges from 10 to 25 percent depending on the method of processing the lamination steel and the method of assembling the magnetic core.

3. The magnetic core loss can also be reduced by using silicon grades of electrical steel. In general, the higher the silicon content up to 4 percent silicon, the lower the magnetic loss in watts per pound. The increase in the silicon content of the lamination steel decreases both the hysteresis and eddy current losses in the magnetic structure. Table 2.2 shows the influence of lamination thickness and grade of electrical steel on the standard epstein losses.

However, because of variables in the processing of the lamination steel into finished motor cores, the reduction in core loss in watts per pound equivalent to the epstein data on flat strips of the lamination steel is seldom achieved. Magnetic core loss reductions on the order of 15 to 40 percent can be achieved by the use of thinner-gauge silicon-grade electrical steels. A disadvantage of the higher-silicon lamination steel is that at high inductions the permeability may be lower, thus increasing the magnetizing current required. This will tend to decrease the motor power factor.

Table 2.2 Typical Epstein Data for Electrical Steels

		Magnetic loss (W/lb at 15 kG)		
Steel grade	Thickness	Hysteresis loss	Eddy current loss	Total loss
Nonsilicon	0.018	1.33	1.42	2.75
	0.024	1.33	2.30	3.63
M-45	0.0185	1.32	0.70	2.02
	0.025	1.32	1.11	2.43
M-36	0.0185	1.19	0.64	1.83
	0.025	1.19	0.87	2.06

Source: Courtesy of U.S. Steel Corporation, Pittsburgh.

Friction and Windage Losses

Friction and windage losses are caused by the friction in the bearings of the motor and the windage loss of the ventilation fan and other rotating elements of the motor. The friction losses in the bearings are a function of bearing size, speed, type of bearing, load, and lubrication used. This loss is relatively fixed for a given design, and since it is a small percentage of the total motor losses, design changes to reduce this loss do not significantly affect the motor efficiency. Most of the windage losses are associated with the ventilation fans and the amount of ventilation required to remove the heat generated by other losses in the motor such as the winding power losses I^2R, magnetic core loss, and stray load loss. As the heat-producing losses are reduced, it is possible to reduce the ventilation required to remove those losses, and thus the windage loss can be reduced. This applies primarily to totally enclosed fan-cooled motors with external ventilation fans. One of the important by-products of decreasing the windage loss is a lower noise level created by the motor.

Stray Load Losses

Stray load losses are residual losses in the motor that are difficult to determine by direct measurement or calculation. These losses are load related and are generally assumed to vary as the square of the output torque. The nature of this loss is very complex. It is a function of many of the elements of the design and the processing of the motor. Some of the elements that influence this loss are the stator winding design, the ratio of air gap length to rotor slot openings, the ratio of the number of rotor slots to stator slots, the air gap flux density, the condition of the stator air gap surface, the condition of the rotor air gap surface, and the bonding or welding of the rotor conductor bars to rotor lamination. By careful design some of the elements that contribute to the stray loss can be minimized. Those stray losses that relate to processing such as surface conditions can be minimized by careful manufacturing process control. Because of the large number of variables that contribute to the stray loss, it is the most difficult loss in the motor to control.

Summary of Loss Distribution

Within a limited range, the various motor losses discussed are independent of each other. However, in trying to make major improvements in efficiency, one finds that the various losses are very dependent. The final motor design is a balance among several losses to obtain a high efficiency and still meet other performance criteria including locked-rotor torque, locked-rotor amperes, breakdown torque, and the power factor.

The distribution of electric motor losses at the rated load is shown in Table 2.3 for several horsepower ratings. It is important for the motor designer to understand this loss distribution in order to make design changes to improve motor efficiency. In a very general sense, the average loss distribution for standard NEMA design B motors can be summarized as follows:

Motor component loss	Total loss (%)
Stator power loss I^2R	37
Rotor power loss I_2^2R	18
Magnetic core loss	20
Friction and windage	9
Stray load loss	16

Table 2.3 Typical Loss Distribution of Standard NEMA Design B Drip-proof Motors[a]

hp		1			5		
Loss distribution	Watts	% Loss[b]	PU loss[c]	Watts	% Loss	PU loss	Watts
Stator power loss I_1^2R	120	43	0.16	305	40	0.08	953
Rotor power loss I_2^2R	35	13	0.05	150	20	0.04	479
Magnetic core loss	76	28	0.10	225	29	0.06	351
Friction and windage loss	24	9	0.03	30	4	0.01	168
Stray load loss	19	7	0.03	51	7	0.01	345
Total losses	274	100	0.37	761	100	0.20	2,296
Output, W	746			3730			18,560
Input, W	1020			4491			20,946
Efficiency, %	73			83			89

[a]Polyphase four-pole motor, 1750 rpm.
[b]% loss = percent of total losses.
[c]PU loss = loss/(hp × 746).

This loss distribution indicates the significance of design changes to increase the electric motor efficiency. However, as the motor efficiency and the horsepower increase, the level of difficulty in improving the electric motor efficiency increases. Consider the stator and rotor power losses only. To improve the motor full-load efficiency, one efficiency point requires an increasing reduction in these power losses as the motor efficiency increases:

hp	Original efficiency (%)	Increased efficiency (%)	Decrease in power losses required (%)
1	73.0	74.0	8
5	83.0	84.0	11
25	89.0	90.0	16
50	90.5	91.5	19
100	91.5	92.5	28
200	93.0	94.0	38

These loss reductions can be achieved by increasing the amount of material, i.e., magnet wire in the stator winding and aluminum conduc-

Table 2.3 *(continued)*

25			50			100			200		
% Loss	PU loss	Watts	% Loss	PU loss	Watts	% Loss	PU loss	Watts	% Loss	PU loss	
42	0.05	1,540	38	0.04	1,955	28	0.026	3,425	30	0.023	
21	0.03	860	22	0.02	1,177	18	0.016	1,850	16	0.012	
15	0.02	765	20	0.02	906	13	0.012	1,650	15	0.011	
7	0.01	300	8	0.01	992	14	0.013	1,072	10	0.007	
15	0.02	452	12	0.01	1,900	27	0.025	3,235	29	0.022	
100	0.13	3,917	100	0.10	6,930	100	0.092	11,232	100	0.075	
		37,300			74,600			149,200			
		41,217			81,530			160,432			
		90.5			91.5			93.0			

tors in the rotor or squirrel-cage winding. However, a loss deduction of only 5 to 15 percent can be achieved in these power losses without making other design modifications. These modifications can include a new lamination design to increase the amount of magnet wire and aluminum rotor conductors that can be used, combined with the use of lower-loss electrical-grade lamination steel in the magnetic structure and the use of a longer magnetic structure. The level of difficulty and consequently the cost of improving the electric motor efficiency increases as the horsepower rating increases. This is illustrated in Fig. 2.2, which shows the decrease in per unit losses as the horsepower rating increases, thus requiring a larger per unit loss reduction at the higher horsepower ratings for the same efficiency improvement.

2.4   WHAT IS AN ENERGY-EFFICIENT MOTOR?

Unfortunately, there is no single definition of an energy-efficient motor. Similarly, there are no efficiency standards for standard NEMA design B polyphase induction motors. As discussed earlier, standard motors were designed with efficiencies high enough to achieve the allowable temperature rise for the rating. Therefore, for a given horsepower rating, there is a considerable variation in efficiency. This is illustrated in Fig. 2.1 for the horsepower range of 1 to 200 hp.

So what is an energy-efficient motor? In 1974, one electric motor manufacturer examined the trend of increasing energy costs and the costs of improving electric motor efficiencies. The cost/benefit ratio at that time justified the development of a line of energy-efficient motors with losses approximately 25 percent lower than the average NEMA design B motors. Based on a 25 percent loss reduction, Fig. 2.3 shows the wattage losses required for energy-efficient motors versus standard motors. Also, Fig. 2.4 illustrates the higher efficiencies of energy-efficient motors based on these criteria.

Subsequent to this development, all major electric motor manufacturers have developed product lines of energy-efficient motors. Since, as previously discussed, there is no standard for the efficiency of motors, the energy-efficient motors of the various manufacturers can generally be identified by their trade names. In addition these products are supported by appropriate published data. Examples of these trade names and their manufacturers are the following:

E-Plus (Gould Inc.)
Energy Saver (General Electric Co.)

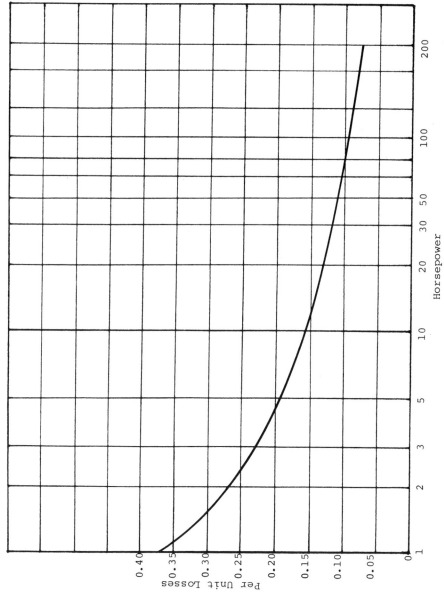

Fig. 2.2 Per unit losses for standard design B four-pole motors.

43

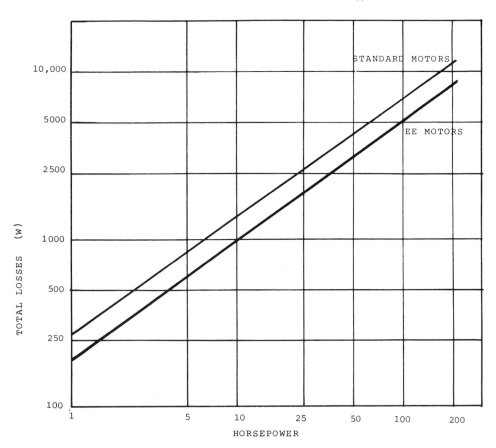

Fig. 2.3 Full-load losses for NEMA design B four-pole motors: standard versus energy-efficient motors (with 25 percent lower losses).

XE—Energy Efficient (Reliance Electric Co.)
Mac II High Efficiency (Westinghouse Electric Corp., Division of Emerson Electric Co.)
Energy Efficient Corro-duty (U.S. Electrical Motors)
High Efficiency Pacemaker (Louis Allis Co.)
Delco E² Motors (Delco Products Division, General Motors)

A survey of the published data available from the manufacturers of energy-efficient motors is summarized in Table 2.4 and Fig. 2.5. These data show the nominal average efficiency as well as the range of nominal efficiencies expected. The efficiencies are shown as nominal

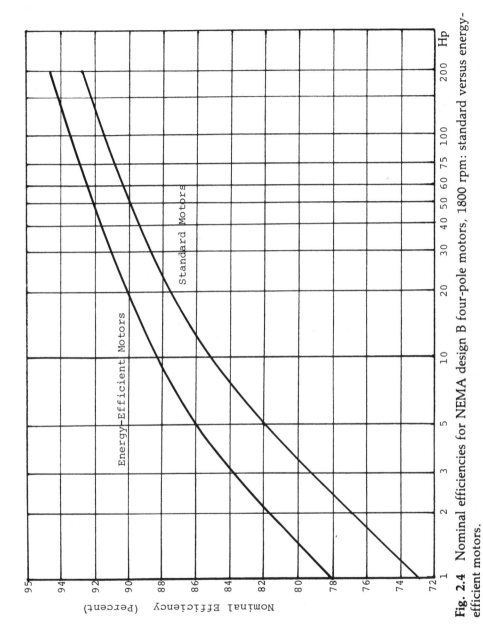

Fig. 2.4 Nominal efficiencies for NEMA design B four-pole motors, 1800 rpm: standard versus energy-efficient motors.

Table 2.4 Full-Load Efficiencies of Three-Phase Four-Pole Energy-Efficient Motors[a]

hp	Nominal efficiency range	Average nominal efficiency
1	80–84	83.0
1.5	81–84	83.0
2	81–84	83.0
3	83.5–88.5	86.0
5	85–88.5	87.0
7.5	86–90.5	88.0
10	87.5–90.5	89.0
15	89.5–91.5	90.0
20	90.0–93.0	90.5
25	91.0–93.0	91.5
30	91.0–93.0	92.0
40	91.5–93.0	92.5
50	91.5–94.0	93.0
60	91.0–94.0	93.0
75	92.0–95.0	93.5
100	93.0–95.0	94.0
125	93.0–95.0	94.0
150	93.0–96.0	94.5
200	94.0–95.5	94.5

[a]Based on available published data.

efficiencies as defined in NEMA Standards Publication MG1. When these efficiency data are compared to the standard motor efficiency data shown in Fig. 2.1, the range in efficiency for a given horsepower is considerably less; in other words, energy-efficient motors tend to be more uniform than standard motors.

When the average nominal efficiency for industry energy-effi-

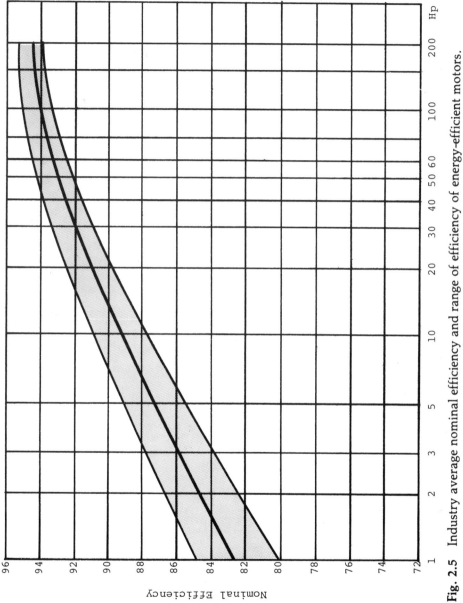

Fig. 2.5 Industry average nominal efficiency and range of efficiency of energy-efficient motors.

47

cient motors shown in Table 2.4 is compared to the data shown in Fig. 2.4 for standard motors, the industry average is consistently higher. When the average efficiency for standard motors in Fig. 2.1 is compared to the average efficiency for energy-efficient motors in Fig. 2.5, the average loss reduction is 32 percent, thus indicating a continuing trend to higher-efficiency motors. These improvements in efficiency, or loss reductions, are generally achieved by increasing the amount of active material used in the motors and by the use of lower-loss magnetic steel. Figure 2.6 shows this comparison of a standard motor and an energy-efficient motor for a particular horsepower rating. In addition to increasing the motor efficiency, there are other user benefits in the application of energy-efficient motors, which will be discussed in more detail in Chap. 5. This trend will probably continue as the cost of power and the demand for higher-efficiency motors continue to increase. Figure 2.7 is a projection of the improvement in the nominal efficiency of the 5- and 50-hp polyphase induction motors. Other horsepower ratings will follow a similar trend.

2.5   EFFICIENCY DETERMINATION

Efficiency is defined as the ratio of the output power to the input power to the motor expressed in percent; thus,

$$\text{Efficiency} = \frac{W_{out} \times 100}{W_{in}}$$

Fig. 2.6   Comparisons of energy-efficient and standard motors. (Courtesy of Gould Inc., Electric Motor Division, St. Louis.)

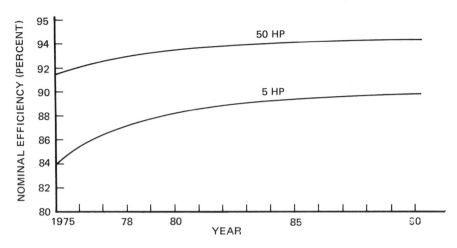

Fig. 2.7   Projected efficiency trends of energy-efficient motors.

It may also be expressed as

$$\text{Efficiency} = \frac{W_{out} \times 100}{W_{out} + W_{loss}}$$

where
   W_{out} = output power, W
   W_{in} = input power, W
   W_{loss} = motor losses, W
The total motor losses include the following losses:

$$W_{loss} = W_s + W_r + W_c + W_f + W_{sl}$$

where
   W_s = stator winding loss
   W_r = rotor winding loss, slip loss
   W_c = magnetic core loss
   W_f = no load friction and windage loss
   W_{sl} = full-load stray load loss
   The accuracy of the efficiency determination depends on the test method used and the accuracy of the losses determined by the test method. There is no single standard method used throughout the industry. The most commonly referred to test methods are the following: *IEEE Standard 112-1978 Standard Test Procedure for Polyphase In-*

duction Motors and Generators; International Electrotechnical Commission (IEC) Publication 34-2, *Methods of Determining Losses and Efficiency of Rotating Electrical Machinery from Tests;* and Japanese Electrotechnical Commission (JEC) Standard 37 (1961), *Standard for Induction Machines.*

Each of these standards allows for more than one method of determining motor efficiency, and they can be grouped into two broad categories: *direct measurement methods* and *segregated loss methods.* In the direct measurement methods, both the input power and output power to the motor are measured directly. In the segregated loss methods, one or both are not measured directly. With direct measurement methods,

$$\text{Efficiency} = \frac{\text{output power}}{\text{input power} \times 100}$$

With segregated loss methods,

$$\text{Efficiency} = \frac{\text{input power} - \text{losses}}{\text{input power}} \times 100$$

or

$$\text{Efficiency} = \frac{\text{output power}}{\text{output power} + \text{losses}} \times 100$$

IEEE Standard 112-1978

Methods A, B, and C are direct measurement methods:

Method A: Brake. In this method a mechanical brake is used to load the motor, and the output power is dissipated in the mechanical brake. The brake's ability to dissipate this power limits this method primarily to smaller sizes of induction motors (generally fractional horsepower).

Method B: Dynamometer. In this method the energy from the motor is transferred to a rotating machine (dynamometer) which acts as a generator to dissipate the power into a load bank. The dynamometer is mounted on a load scale, a strain gauge, or a torque table. This is a very flexible and accurate test method for motors in the range from 1 to 500 hp. However, to ensure accuracy, dynamometer corrections should be made as outlined in the test procedure. Method B modified to include stray loss data smoothing by a linear regression analysis has been adopted by NEMA as the method for determining the efficiency of motors in this horsepower range.

Method C: Duplicate Machines.   This method uses two identical motors mechanically coupled together and electrically connected to two sources of power, the frequency of one being adjustable.*
Readings are taken on both machines, and computations are made to calculate efficiency.

Methods E and F are segregated loss methods:

Method E: Input Measurements.†   The motor output power is determined by subtracting the losses from the measure motor input power at different load points. For each load, the measured I^2R losses are adjusted for temperature and added to the no load losses of friction, windage, and core. The stray load loss, which may be determined either directly, indirectly, or by the use of an agreed upon standardized value, is included in this total.

Method F: Equivalent Circuit Calculations.   When load tests cannot be made, operating characteristics can be calculated from no load and impedance data by means of an equivalent circuit. This equivalent circuit is shown in Fig. 2.8. Because of the nonlinear nature of these circuit parameters, they must be determined with great care to ensure accurate results. Procedures for determining these parameters are outlined in the standard as determined by a separate test.

IEC Publication 34-2

The same basic alternate methods as outlined for IEEE 112 are also allowed for in IEC 34-2. However, a preference is expressed for the sum-

---

*One machine is operated as a motor at rated voltage and frequency, and the other is driven as a generator at rated voltage per hertz but at a lower frequency to produce the desired load.
†In this method it is necessary to connect the motor to a variable load. The input power is measured at the desired load points.

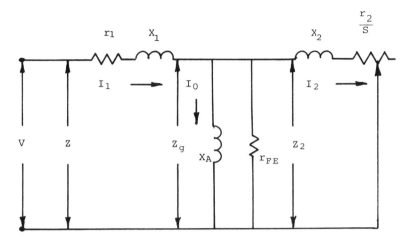

Fig. 2.8 Polyphase induction motor per phase equivalent circuit. (From R. E. Osterlei, *Proceedings of the 7th National Conference on Power Transmission*, Gould Inc., St. Louis, 1980.)

mation of losses method for the determination of motor efficiency. This is similar to IEEE 112 methods E and F, except that the IEC method specifies stray load loss and temperature corrections differently. The IEC stray load losses are assumed to be 0.5 percent of rated input, whereas the IEEE standard states a preference for direct measurement of the stray load losses. The resistance temperature corrections in the IEC method are given as fixed values depending on insulation class, while the IEEE standard recommends use of the measured temperature rise for correcting resistance. These differences generally result in higher motor efficiency values by the IEC method.

JEC Standard 37

The JEC 37 standard also specifies the same basic methods as IEEE 112 with the exception of method C, duplicate machines. The preferred method for determining efficiency in this standard is by the use of circle diagrams. This is a graphical solution of the T equivalent circuit of the induction motor. (This is similar to IEEE method F without the R_{fe} circuit branch.) As in the IEC standard, different methods are used to determine the circuit parameters and adjust the performance calculations. Principal among these is the setting of stray load loss equal to zero and using fixed values for resistance temperature corrections which are a function of insulation class. These differences generally produce higher values for motor efficiency than the IEEE methods.

Comparison of Efficiencies Determined by Preferred Methods

To illustrate the variations in efficiency resulting from the use of the preferred methods, the full-load efficiency of several different polyphase motors was calculated by the preferred test methods given in the three standards. The results are shown in Table 2.5. As the values show, the efficiencies determined by the IEC and JEC methods are higher than the IEEE method. The major reason for this difference is the way in which stray load losses are accounted for. The IEEE method B stray load losses are included in the direct input and output measurements, whereas in the IEC method the stray load losses are taken as 0.5 percent of the input, and in the JEC method they are set equal to zero. This comparison shows how important it is to know the method used to determine efficiency when comparing electric motor performance from different sources and countries.

Testing Variance

In addition to variance in efficiency due to test methods used, variances can also be caused by human errors and test equipment accuracy. With dynamometer (IEEE 112, method B) testing, as with all test methods, there are several potential sources of inaccuracies: in-

Table 2.5 Efficiency Determined by Preferred Methods

hp	JEC 37, circle diagram	IEC 34-2 loss summation	IEEE 112, method B
5	88.8	88.3	86.2
10	89.7	89.2	86.9
20	91.9	91.4	90.4
75	93.1	92.7	90.0

Source: R. E. Osterlei, *Proceedings of the 7th National Conference on Power Transmission,* Gould Inc., 1980.

Table 2.6   Variation in Test Data

	Variation in full-load efficiency			
	Without stray smoothing		With stray smoothing	
hp rating	Mean efficiency	Variation ±two std. deviations	Mean efficiency	Variation ±two std. deviations
5	86.3	2.0	87.1	0.7
25	89.6	1.3	89.5	0.8
100	92.7	1.3	91.9	0.9

Source: R. E. Osterlei, *Proceedings of the 7th National Confer-ence on Power Transmission,* Gould Inc., St. Louis, 1980.

strument accuracy, dynamometer accuracy, and instrument and dy-namometer calibration. Therefore, to minimize these test errors, it is recommended that all the equipment and instruments be calibrated on a regular basis.

With proper calibration, dynamometer testing provides consis-tent and verifiable electric motor performance comparison. NEMA conducted a round-robin test of three different horsepower ratings (5, 25, and 100 hp) with a number of electric motor manufacturers. After a preliminary round of testing, each manufacturer was requested to test the motors in accordance with IEEE 112, method B, both with and without mathematical smoothing of the stray load loss. The results of these tests are summarized in Table 2.6.

Based on the test results, NEMA adopted a standard test proce-dure for polyphase motors rated 1 to 125 hp in accordance with IEEE Standard 112, method B, including mathematical smoothing of the stray load loss. It is recommended that this method of determining motor efficiency be used wherever possible.

2.6   MOTOR EFFICIENCY LABELING

Coincident with the NEMA test program, it was determined that a more consistent and meaningful method of expressing electric motor efficiency was necessary. The method should recognize that motors, like any other product, are subject to variations in material, manufacturing processes, and testing that cause variations in efficiency on a motor-to-motor basis for a given design.

No two identical units will perform in exactly the same way. Variance in the electrical steel used for laminations in the stator and rotor cores will cause variance in the magnetic core loss. Variance in the diameter and conductivity of the magnet wire used in the stator winding will change the stator winding resistance and hence the stator winding loss. Variances in the conductivity of aluminum and the quality of the rotor die casting will cause changes in the rotor power loss. Variances also occur in the manufacturing process. The quality of the heat treatment of the laminations for the stator and rotor cores can vary, causing a variance in the magnetic core loss. The winding equipment used to install the magnet wire in the stator can have tension that is too high, stretching the magnet wire and thus increasing the stator winding resistance and resulting in an increase in the stator winding loss I^2R. Sim-

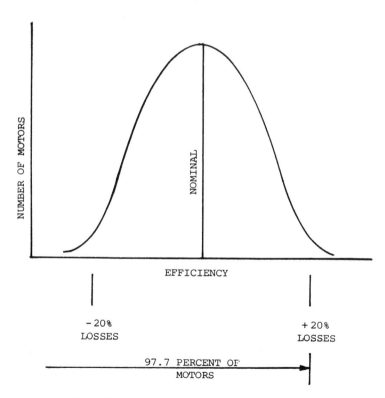

Fig. 2.9 Normal frequency distribution.

ilarly, other variances, such as dimensional variances of motor parts, will contribute to the variation in motor efficiency.

It is a statistical fact that a characteristic of a population of a product will generally be distributed according to a bell-shaped or Gaussian distribution curve. The height of the curve at any point is proportional to the frequency of occurrences, as illustrated in Fig. 2.9.

In the case of electric motors, the variation of losses for a population of motors of a given design is such that 97.7 percent of the motors will have an efficiency above the minimum efficiency defined by a variation of motor losses of ±20 percent of the losses at the nominal or average efficiency. Figure 2.10 illustrates the efficiency distribution for a specific value of nominal efficiency of 91 percent.

It is possible as motor manufacturers gain experience with this procedure that variations in losses will be lower than ±20 percent. In this event, the spread between nominal efficiency and minimum efficiency can be reduced.

Consequently, NEMA adopted a standard publication, MG1-12.53b, recommending that polyphase induction motors be labeled

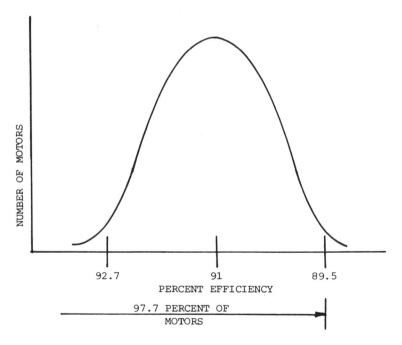

Fig. 2.10 Normal efficiency distribution.

Table 2.7   NEMA Nominal Efficiencies

Nominal efficiency	Minimum efficiency	Nominal efficiency	Minimum efficiency
95.0	94.1	81.5	78.5
94.5	93.6	80.0	77.0
94.1	93.0	78.5	75.0
93.6	92.4	77.0	74.0
93.0	91.7	75.5	72.0
92.4	91.0	74.0	70.0
91.7	90.2	72.0	68.0
90.0	89.5	70.0	66.0
90.2	88.5	68.0	64.0
89.5	87.5	66.0	62.0
88.5	86.5	64.0	59.5
87.5	85.5	62.0	57.5
86.5	84.0	59.5	55.0
85.5	82.5	57.5	52.5
84.0	81.5	55.0	50.5
82.5	80.0	52.5	48.0
		50.5	46.0

Source: Reprinted by permission from NEMA Standards Publication No. MG1-1978, *Motors and Generators,* copyright 1978 by the National Electrical Manufacturers Association.

with a *NEMA nominal efficiency* (or *NEMA NOM EFF*) when tested in accordance with IEEE Standard 112, dynamometer method, with stray loss smoothing. In addition, a minimum efficiency value was developed for each nominal efficiency value. Table 2.7 is a copy of the NEMA efficiency table. It is recommended that this method of labeling efficiency and testing be specified whenever possible.

In instances where guaranteed efficiencies are required, it is recommended that the preceding test method or an appropriate test method including the method of loss determination and the losses to be included in the efficiency determination be specified.

3

Electric Power Costs

3.1 INDUSTRY ENERGY TRENDS

The industrial sector of the U.S. economy is the largest consumer of energy in the economy. As a consumer of a wide variety of fuels, the industrial sector is a focal point for government policies designed to reduce the nation's reliance on imported oil. The effect of these policies and of higher oil prices on the industrial fuel mix has been a reversal of historical trends of energy consumption. The overall energy intensity of industrial activities (measured by the Btu consumed per constant dollar of manufacturing value added) is projected to continue to decline, as shown in Fig. 3.1. However, to continue this trend, industry must search for and utilize energy saving technologies and products. Energy-efficient electric motors are one of the energy-saving products that can contribute to this trend.

3.2 ELECTRIC POWER COST TRENDS

Through the 1960s into the early 1970s, the cost of electric power was relatively constant. With the advent of the oil price escalation starting in the early 1970s, the cost of electric power in the industrial sector has been increasing at an annual rate of 11 to 12 percent. Figure 3.2 shows the relative change in industrial electric power rates compared to the consumer price index (CPI) from 1965 to 1979. The projections are that this trend will continue. The U.S. Department of Energy in their 1979 annual report to Congress [DOE/EIA-0173(79)/3] made projections of

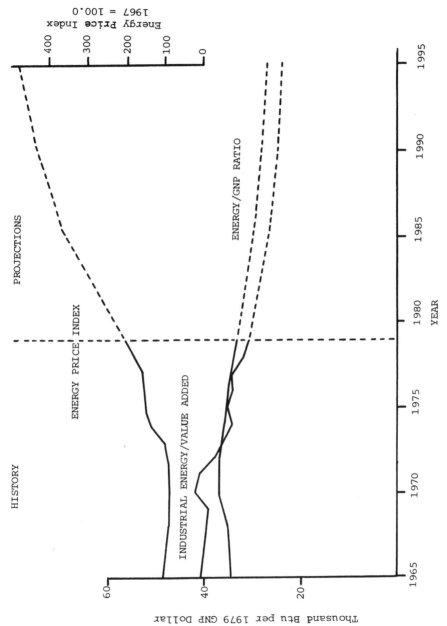

Fig. 3.1 U.S. energy and economic indicators. (From DOE/EIA-0173(79)/3.)

59

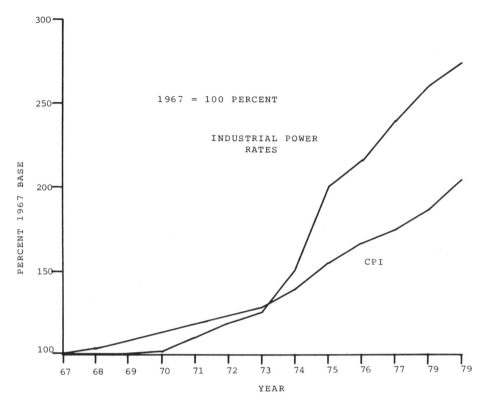

Fig. 3.2 Relative cost comparison of the consumer price index and industrial power rates. (From DOE/EIA-0040(79).)

energy supply and costs under selected assumptions. As pointed out in the report, several sources of uncertainty exist and include the following:

The price and availability of imports
The size of the domestic resource base
The health of the total economy
The rate at which capital stock changes in each sector
Changes in government policy

These projected economic variables and industrial electrical energy costs through 1995 are shown in Table 3.1. Note that these data are in constant 1979 dollars and so do not reflect inflation rates. A comparison of the projection of the consumer price index and electric power costs through 1990 is shown in Fig. 3.3. It is important to recognize that these projections are subject to the uncertainties previously outlined. However, they indicate a continuing increase in the cost of industrial electric power at a higher rate than the CPI.

3.3   ELECTRIC UTILITY RATES

The rate structures of the electric utilities are complex documents. Because industrial power contracts contain so many variables, it is not likely that two plants with the same mode of operation could be served by any two utilities in the country and both pay the same electric bill. This will be illustrated later in this chapter. In addition, almost every electrical utility is involved in some form of rate reform. Utilities are faced with a financial crunch due to operating costs and construction costs, so many are looking for ways to revamp outmoded formulas used to compute electric rates. It is therefore of paramount importance that the plant engineer understand not only the existing rate structures but also pending rate reforms and their effects on power bills.

The electric power rates and escalation of these rates depends to some degree on the location and the utility serving that location. Table 3.2 shows the weighted average bills by geographic division. For 1979 this table shows a range in electric power costs of 20 percent below to 45 percent above the average.

It is essential for the plant engineer to understand all the elements that are included in his or her plant's electric power costs: demand charge, power factor charge, energy charge, fuel adjustment charge, equipment rental, special use charges, use or franchise taxes, and sales taxes.

Demand Charge

The demand charge is based on the maximum kilowatt demand per month. The time interval for determining the kilowatt demand is usually 15 or 30 min. Depending on the specific rate structure, there may be other requirements included in determining the demand charge such as average demand for the preceding 11 months, or a percent (such as 75 percent) of the contracted demand, or a percent (such as 85 percent) of the maximum summer demand in the preceding 11 months. In addition the demand charge for the summer months may be substantially higher than that for the winter months.

Table 3.1   Selected Macroeconomic Variables: 1965–1979 Historical and Final Values for Three Base Scenario Projections, 1985–1995

	History			
World oil price (1979 dollars per barrel):	1965, 6.00	1973, 6.50	1978, 15.50	1979, NA
Macroeconomic variables				
Real gross national product, billions of 1979 dollars	1533	2044	2314	2369
Compound annual rate of growth in real GNP, to/from 1978	3.2	2.5	NA	2.3
Production index for manufacturing, 1967 = 1.00	0.90	1.30	1.47	1.53
Implicit price deflator for GNP, 1972 = 1.00	0.74	1.06	1.47	1.66
Rate of increase in the consumer price index, to/from 1978	5.8	8.0	NA	11.3
Energy variables				
Wholesale price index, fuels and related products, and power, 1967 = 1.00	0.95	1.34	3.23	4.08
Energy/GNP ratio, 1000 Btu/ 1979 GNP dollar	34.6	36.4	33.7	32.9
Electrical energy				
Industrial electricity, ¢/kWh	2.20	2.00	2.90	3.1

Source: DOE/EIA-0173(79)/3.

Table 3.1 *(continued)*

	Projections							
1985			1990			1995		
Low, 27.00	Mid, 32.00	High, 39.00	Low, 27.00	Mid, 37.00	High, 44.00	Low, 27.00	Mid, 41.00	High, 56.00
2734	2718	2696	3209	3159	3116	3650	3569	3501
2.4	2.3	2.2	2.8	2.6	2.5	2.7	2.6	2.5
1.87	1.86	1.84	2.34	2.28	2.24	2.72	2.66	2.61
2.61	2.64	2.69	3.55	3.64	3.70	4.65	4.73	4.81
8.6	8.8	9.2	7.8	8.1	8.3	7.4	7.5	7.7
10.8	12.2	14.2	15.9	19.9	23.1	21.7	28.9	37.0
30.4	29.9	29.9	28.8	28.2	28.1	27.7	27.0	26.9
3.80	3.90	3.90	4.10	4.20	4.20	4.00	4.10	4.10

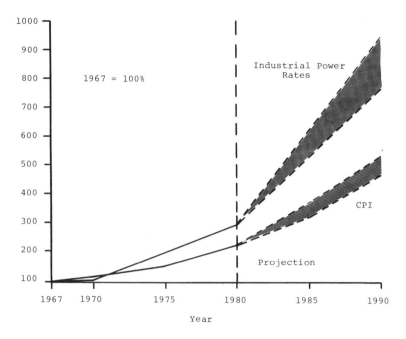

Fig. 3.3 Relative cost projections of the consumer price index and industrial power rates. (From DOE/EIA-0040(79) and DOE/EIA-0173(79)/3.)

In other instances, there is a different demand charge rate for off-peak hours. This off-peak demand charge rate is generally substantially lower than the peak hour demand charge rate. It is an item not to be overlooked in controlling the plant demand charge.

In many cases, the demand is also reflected in the energy rate, which is set on the basis of kilowatt-hours per kilowatt demand. In those instances, the lower the demand, the higher the kilowatt-hours per kilowatt demand and the sooner the energy cost is reduced to the next increment on the energy rate schedule.

Therefore, the demand charge can represent a significant share of the electric power bill, and it is usually economical to determine methods to reduce the kilowatt demand.

The principle of demand control is fairly simple: It is necessary to determine at what time of day and on which days the peak demand occurs and then which loads are in use at the peak demand condition. The problem of demand control now becomes quite complex. Which of these loads can be deferred or curtailed to reduce the peak demand?

Table 3.2 Weighted Average Bills for Industrial Service by Geographic Location (Large Cities Only)

Geographic location	January 1, 1979			January 1, 1978			January 1, 1969		
	150 kW, 30,000 kWh	300 kW, 60,000 kWh	1000 kW, 200,000 kWh	150 kW, 30,000 kWh	300 kW, 60,000 kWh	1000 kW, 200,000 kWh	150 kW, 30,000 kWh	300 kW, 60,000 kWh	1000 kW, 200,000 kWh
New England	$1459	$2794	$ 8,851	$1502	$2887	$ 9,185	$675	$1253	$3778
Middle Atlantic	2141	4246	13,621	2107	4104	13,382	811	1490	4464
East north central	1521	2904	9,144	1364	2620	8,188	667	1212	3513
West north central	1307	2560	8,074	1222	2389	7,581	641	1177	3460
South Atlantic	1409	2690	8,498	1381	2596	8,358	561	1013	3921
East south central	1226	2390	7,432	1109	2154	6,641	483	893	2597
West south central	1246	2398	7,467	1147	2187	6,615	553	999	2831
Mountain	1350	2578	7,926	1,253	2387	7,334	577	1078	3199
Pacific	1287	2525	7,978	1301	2474	7,878	448	818	2446
Noncontiguous	1673	3220	10,110	1592	3058	9,562	696	1292	3436
U.S. average	1531	2975	9,408	1469	2824	8,973	636	1163	3436
Average cost, ¢/kWh	5.10	4.96	4.70	4.90	4.71	4.49	2.12	1.94	1.72

Source: DOE/EIA-0400(79).

The decision on how to limit the peak demand depends on the nature of the loads and the complexity of the system. This can range from simple to very complex controls:

Manual control of selected loads
Time switches
Interlocks
Automatic demand controllers
Microcomputer load management systems

Any demand control equipment selected should have enough flexibility to be adjusted to changes in the utility rate structure and maintain economical power costs.

Power Factor Charge

The power factor charge is a separate charge or is reflected in an adjustment of the demand charge for the system power factor below a specified level.

The power factor charge may be a separate charge, such as $0.20/kvar peak registered on a reactive power meter, but more likely it is an adjustment to the demand charge. Typical demand charge adjustments include the following:

1. For a power factor over 0.90, the demand charge is decreased by 2 percent for each point of the power factor in excess of 0.90. For a power factor below 0.80, the demand charge is increased by a factor of 0.80/actual PF.
2. The billing demand is multiplied by a factor of 0.80 + (0.6 kvar/kW).
3. If the power factor falls below 0.85, the demand charge and all energy usage charges over 300 kWh will be increased by a factor of 1.111.

Some users may think they are not being penalized on paying for a low power factor since there is no specified power factor charge. In those cases, the billing demand may be based on kilovolt-ampere demand instead of kilowatt demand. For example, if a plant is operating with a kilowatt demand at a power factor of 0.75 instead of 0.85, the kilovolt-ampere demand charge would be 13 percent higher for the same kilowatt demand. Additional charges would be reflected in the energy charge if the rate structure is a function of the kilowatt-hours of usage per kilovolt-ampere of demand.

Energy Consumption Charge

The energy consumption charge is that portion of the electric utility bill based on the kilowatt-hours of electric energy consumed during the billing period.

In many instances, the utility rate structure energy charge is a declining block rate based on kilowatt-hour usage, for example,

First 50 kWh: $3.60 flat charge
Next 450 kWh: $0.0621/kWh
Next 14,500 kWh: $0.0521/kWh
Over 15,000 kWh: $0.0231/kWh

These rates may be different for summer and winter, with the summer rates being as much as 50 percent higher than the winter rates.

In other cases, the rate structure is more complex and is based on the kilowatt-hours of consumption per kilowatt of demand or the kilowatt-hours of consumption per kilovolt-ampere of demand, for example,

First 100 kWh/kW of demand: $0.0183/kWh
Next 120 kWh/kW of demand: $0.0167/kWh
Next 110 kWh/kW of demand: $0.0150/kWh
Over 440 kWh/kW of demand: $0.0130/kWh

In a few instances the energy charge may be a flat rate per kilowatt-hour of consumption.

Again, the plant engineer should be alert to rate differentials that may exist between peak and off-peak hours in order to take advantage of off-peak hour rates. Other special rate schedules should be reviewed such as the water-heating service, electric space heating, and interruptible service.

Fuel Adjustment Charge

In addition to the energy consumption charge, many utilities have a purchased power and fuel adjustment charge. This charge is generally imposed on the basis of kilowatt-hours of consumption for the billing period.

The method of application of this charge depends on the ruling of each state public utility regulating agency, and it is usually stated as a separate item on the utility bill.

This charge is usually adjusted monthly or quarterly based on a prescribed formula relating to the average delivered cost of fuel or purchased power. The following is an example of this type of charge: "The fuel cost adjustment shall consist of an increase or decrease of 0.0109 mills/kWh for each full 0.01-mill increase or decrease in the average delivered cost of fuel burned monthly above or below a base cost of 18.36 mills/kWh."

The range of this charge is from 0 to 7¢/kWh. The current average fuel adjustment charge is on the order of 1.5 to 2¢/kWh. Many utilities do not have a fuel adjustment charge as a stated item in their billing; however, they are obtaining adjustments in their base energy rates every 6 months to compensate for fuel cost increases.

Equipment Rental

Many utilities have nominal charges for the use of utility-owned equipment. This can vary from demand metering equipment to service transformers when the power is metered at the primary power level. This charge is usually stated as a monthly rental charge on the customer billing. In many cases, this charge is a bargain. In other cases it is not, and alternatives should be examined.

3.4 SUMMARY

Electric power rates will continue to increase. Figure 3.4 shows the percentage rate increases from January 1, 1980, to January 1, 1981, for 76

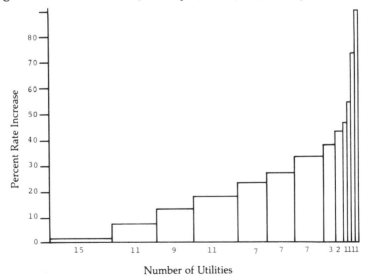

Fig. 3.4 Typical electric utility industrial rate increases from January 1980 to January 1981. (From the U.S. Department of Energy.)

Table 3.3 Electric Energy Cost Comparison for Nine Utilities

Cost feature	Utility								
	1	2	3	4	5	6	7	8	9
Demand period, min	15	30	15	15	30	15	30	30	15
Demand charge, $	89	2489	300	699	1572	856	1602	240	1442
Power factor charge, $	—	—	—	45	—	—	158	15	—
Energy charge, $	4170	1477	2139	2249	2581	1631	910	2890	1870
Fuel adjustment charge, $	451	—	6072	30	760	1556	1360	—	4250
Other charges, $	10	26	10	15	2	6	—	9	243
Rental, $	—	36	—	—	25	—	—	—	—
Taxes, $	188	623	156	—	268	—	—	—	—
Total, $	4908	4651	8377	3038	5208	4049	4030	3154	7805
Cost per kilowatt-hour, ¢	5.77	5.47	9.85	3.57	6.13	4.76	4.74	3.71	9.18

utilities located in 48 states. The figure shows that 50 percent of the rate increases were in the range of a 15 to 45 percent increase. The average of the rate increases was 20 percent.

It is increasingly important to understand the rate structure and anticipated changes of your serving electric utility. As mentioned earlier, no two utilities have the same rate structure, and this variation is illustrated by Table 3.3, which shows the cost of electrical energy for nine utilities for the same demand and energy consumption of 300 kW of demand at an 80 percent power factor and 85,000 kWh. However, the rate elements that make up the total electrical charge exhibit even wider variations from one utility to another. Again, this emphasizes the importance of understanding your electric utility rate structure and the characteristics of your system load so actions can be taken to reduce both the demand charges and energy charges. This knowledge is also necessary to evaluate the additions of new electrical loads and the economics of energy-conserving products such as energy-efficient electric motors.

The conclusion can be drawn that some form of load management system can be justified for most industrial electrical systems. The level of load management equipment required is a function of the size and complexity of the particular electrical system.

4

The Power Factor

The advantages of improving the equipment and system power factor are not as obvious as those of improving the kilowatt power consumption.

4.1 WHAT IS THE POWER FACTOR?

The line current drawn by induction motors, transformers, and other inductive devices consists of two components: the magnetizing current and the power-producing current.

The magnetizing current is that current required to produce the magnetic flux in the machine. This component of current creates a reactive power requirement which is measured in kilovolt-amperes reactive (kilovars, kvar). The power-producing current is that current which reacts with the magnetic flux to produce the output torque of the machine and to satisfy the equation

$$\text{Torque} = K\Phi I$$

where

T = output torque
Φ = net flux in the air gap as a result of the magnetizing current
I = power producing current
K = output coefficient for a particular machine

This component of current creates the load power requirement measured in kilowatts (kW). The magnetizing current and magnetic flux are relatively constant at constant voltage. However, the power-producing current is proportional to the load torque required.

The total line current drawn by an induction motor is the vector sum of the magnetizing current and the power producing current. For three-phase motors, the apparent power or kVA input to the motor is

$$kVA = I_L V_L \sqrt{3} \div 1000$$

where

I_L = total line current
V_L = line-to-line voltage

The vector relationship between the line current I_L and the reactive component I_x and load component I_p currents can be expressed by a vector diagram, as shown in Fig. 4.1, where the line current I_L is the vector sum of two components. The power factor is then the cosine of the electrical angle θ between the line current and phase voltage.

This vector relationship can also be expressed in terms of the components of the total kVA input, as shown in Fig. 4.2. Again the power factor is the cosine of the angle θ between the total kVA and kW inputs to the motor. The kVA input to the motor consists of two components: load power, i.e., kilowatts, and reactive power, i.e., kilovars.

The system power factor can be determined by a power factor meter reading or by the input power (kW), line voltage, and line current readings. Thus,

$$\text{Power factor} = \frac{kW}{kVA}$$

where

$$kVA = I_L V_L \sqrt{3} \div 1000$$

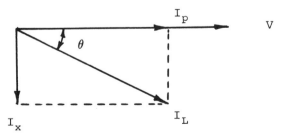

Fig. 4.1   Vector diagram of load current for one-phase of the motor.

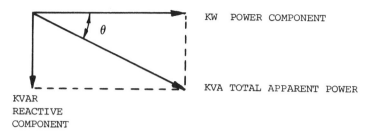

KW   POWER COMPONENT

KVA TOTAL APPARENT POWER

KVAR
REACTIVE
COMPONENT

Fig. 4.2   Vector diagram of power input without a power factor correction.

Then kvar is

$$kvar = \sqrt{(kVA)^2 - (kW)^2}$$

An inspection of the kVA input diagram shows that the larger the reactive kvar, the lower the power factor and the larger the kVA for a given kW input.

4.2   WHY RAISE THE POWER FACTOR?

A low power factor cases poor system efficiency. The total apparent power must be supplied by the electric utility. With a low power factor, or a high kvar component, additional generating losses occur throughout the system. Figures 4.3 and 4.4 illustrate the effect of the power factor on generator and transformer capacity. To discourage low-power-factor loads, most utilities impose some form of penalty or charge in their electric power rate structure for a low power factor.

When the power factor is improved by installing power capacitors or synchronous motors, several savings are made:

1.   A high power factor eliminates the utility penalty charge. This charge may be a separate charge for a low power factor or an adjustment to the kilowatt demand charge.
2.   A high power factor reduces the load on transformers and distribution equipment.
3.   A high power factor decreases the I^2R losses in transformers, distribution cable, and other equipment, resulting in a direct saving of kilowatt-hour power consumption.
4.   A high power factor will help stabilize the system voltage.

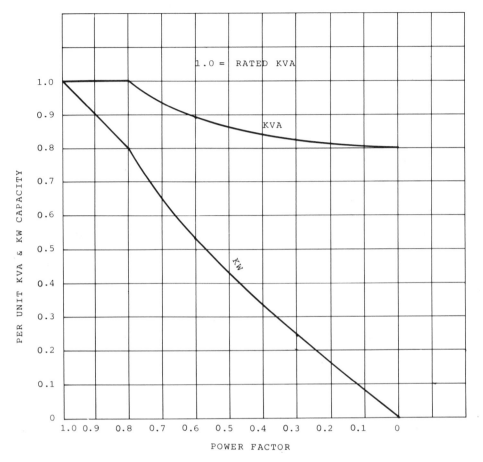

Fig. 4.3   Power factor effect on generator capacity.

4.3   HOW TO IMPROVE THE POWER FACTOR

To improve the power factor for a given load, the reactive load component (kvar) must be reduced. This component of reactive power lags
the power component (kW input) by 90 electrical degrees, so one way
to reduce the effect of this component is to introduce a reactive power
component that leads the power component by 90 electrical degrees.
This can be accomplished by the use of a power capacitor, as illustrated in the power diagram in Fig. 4.5, resulting in a net decrease in the
angle θ or an increase in the power factor.

There are several methods that are used to improve the power

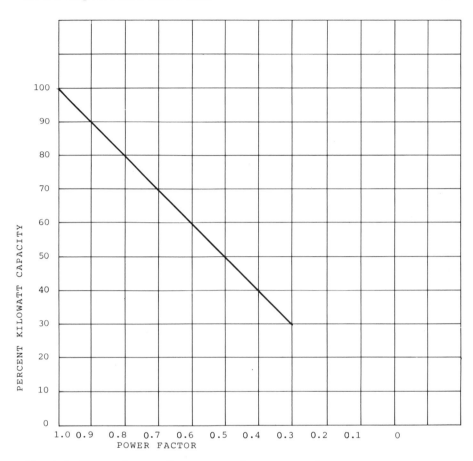

Fig. 4.4 Power factor effect on transformer capacity.

factor in a system installation. One method which can be employed in large systems is to use synchronous motors to drive low-speed loads that require continuous operation. A typical application for a synchronous motor is for driving a low-speed air compressor which provides process compressed air for the plant. The synchronous motor is adjusted to operate at a leading power factor and thus provide leading kvars to offset the lagging kvar of inductive-type loads such as induction motors.

Synchronous motors are usually designed to operate at an 80 percent leading power factor and draw current which leads the line voltage rather than lags it, as is the case with induction motors and transform-

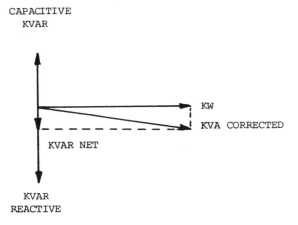

CAPACITIVE
KVAR

KVAR NET

KVAR
REACTIVE

KW

KVA CORRECTED

Fig. 4.5 Vector diagram of power input with a power factor correction.

ers. For example, consider a load of 2000 kW at a 70 percent lagging power factor. The utilization of a 200-hp synchronous motor operating at an 80 percent leading power factor will increase the overall system power factor from 70 percent lagging to 85 percent lagging.

The more popular method of improving the power factor on low-voltage distribution systems is to use power capacitors to supply the leading reactive power required. The amount and location of the corrective capacitance must be determined from a survey of the distribution system and the source of the low-power-factor loads. In addition, the total initial cost and payback time of the capacitor installation must be considered.

To reduce the system losses, the power factor correction capacitors should be electrically located as close to the low-power-factor loads as possible. In some cases, the capacitors can be located at a particular power feeder. In other cases, with large horsepower motors, the capacitors can be connected as close to the motor terminals as possible. The power factor capacitors are connected across the power lines in parallel with the low-power-factor load.

The amount of kvars of capacitors required depends on the power factor without correction and the desired corrected value of the power factor.

The power factor and kvar without correction can be determined by measuring the power factor, line amperes, and line voltage at the point of correction. For a three-phase system,

$$kVA\ input\ =\ \frac{line\ amperes\ \times\ line\ volts\ \times\ \sqrt{3}}{1000}$$

$$kW\ =\ kVA\ \times\ PF$$

$$kvar\ =\ \sqrt{(kVA)^2\ -\ (kW)^2}$$

or the line kilowatts, line amperes, and line voltage can be measured. Then

$$kVA\ input\ =\ \frac{line\ amperes\ \times\ line\ volts\ \times\ \sqrt{3}}{1000}$$

$$PF\ =\ \frac{kW}{kVA}$$

$$kvar\ =\ \sqrt{(kVA)^2\ -\ (kW)^2}$$

The capacitive kvar required to raise the system to the desired power factor can be calculated as follows:

$$kvar\ capacitance\ =\ kvar\ load\ -\ \sqrt{\frac{1\ -\ PF^2}{PF^2}(kW\ load)^2}$$

where PF is the desired power factor.

For example, consider a 1000-kW load with a 58 percent power factor that one wishes to correct to 90 percent:

$$kW\ load\ =\ 1000$$

$$kvar\ load\ =\ \sqrt{\left(\frac{1000}{0.58}\right)^2 -\ (1000)^2}\ =\ 1404\ kvar$$

$$kvar\ capacitance\ =\ 1404\ -\ \sqrt{\frac{1\ -\ (0.09)^2}{(0.90)^2}\ \times\ (1000)^2}\ =\ 920$$

Tables such as Table 4.1 have been developed and are available from most power capacitor manufacturers to simplify this calculation. Table 4.1 provides a multiplier to be applied to the kW load to determine the capacitive kvar required to obtain the desired corrected power factor. Consider the same 1000-kW load with a 58 percent power factor that one wishes to correct to 90 percent. From Table 4.1, for the existing power factor (58 percent) and the corrected power factor (90 percent), the power factor correction factor is 0.919. Thus the kvar of ca-

Table 4.1   Power Factor Correction Factors

Existing power factor (%)	Corrected power factor				
	100%	95%	90%	85%	80%
50	1.732	1.403	1.247	1.112	0.982
52	1.643	1.314	1.158	1.023	0.893
54	1.558	1.229	1.073	0.938	0.808
56	1.479	1.150	0.994	0.859	0.799
58	1.404	1.075	0.919	0.784	0.654
60	1.333	1.004	0.848	0.713	0.583
62	1.265	0.936	0.780	0.645	0.515
64	1.201	0.872	0.716	0.581	0.451
66	1.139	0.810	0.654	0.519	0.389
68	1.078	0.749	0.593	0.458	0.328
70	1.020	0.691	0.535	0.400	0.270
72	0.964	0.635	0.479	0.344	0.214
74	0.909	0.580	0.424	0.289	0.195
76	0.855	0.526	0.370	0.235	0.105
78	0.802	0.473	0.317	0.182	0.052
80	0.750	0.421	0.265	0.130	—
82	0.698	0.369	0.213	0.078	—
84	0.646	0.317	0.161	—	—
86	0.594	0.265	0.109	—	—
88	0.540	0.211	0.055	—	—
90	0.485	0.156	—	—	—

pacitance required is 1000 × 0.919 = 919 kvars.
    Let us verify this calculation:

$$\text{Uncorrected kVA} = \frac{1000}{0.58} = 1724$$

$$\text{Uncorrected lagging kvar} = \sqrt{(1724)^2 - (1000)^2} = 1404$$

$$\text{Correction capacitor kvar} = \underline{919}$$

$$\text{Net corrected lagging kvar} = 485$$

$$\text{Corrected kVA} = \sqrt{(1000)^2 + (485)^2} = 1111$$

$$\text{Corrected power factor} = \frac{1000}{1111} = 0.90$$

In the application of power factor correction capacitors at the motor location, NEMA recommends the following procedure based on the published or nameplate data for the electric motor:

1. The approximately full-load power factor can be calculated from published or nameplate data as follows:

$$PF = \frac{431 \times hp}{E \times I \times Eff}$$

where

PF = per unit power factor at full load (per unit PF = percent PF/100)
hp = rated horsepower
E = rated voltage
I = rated current
Eff = per unit nominal full-load efficiency from published data or as marked on the motor nameplate (per unit Eff = percent Eff/100)

2. For safety reasons, it is generally better to improve the power factor for multiple loads as a part of the plant distribution system. In those cases where local codes or other circumstances require improving the power factor of an individual motor, the kvar rating of the improvement capacitor can be calculated as follows:

$$kvar = \frac{0.746 \times hp}{Eff} \left(\frac{\sqrt{1 - PF^2}}{PF} - \frac{\sqrt{1 - PF_i^2}}{PF_i} \right)$$

where

kvar = rating of a three-phase power factor improvement capacitor
PF_i = improved per unit power factor for the motor-capacitor combination

3. In some cases, it may be desirable to determine the resultant power factor PF_i, where the power factor improvement capacitor selected within the maximum safe value specified by the motor manufacturer is known. The resultant full-load power factor PF_i can be calculated from the following:

$$PF_i = \frac{1}{\sqrt{\{[(\sqrt{1 - PF^2})/PF) - [(kvar \times Eff)/(0.746 \times hp)]\}^2 + 1}}$$

Warning: In no case should power factor improvement capacitors be applied in ratings exceeding the maximum safe value specified by the motor manufacturer. Excessive improvement may cause overexcitation resulting in high transient voltages, currents, and torques that can increase safety hazards to personnel and cause possible damage to the motor or to the driven equipment. For additional information on safety considerations in the application of power factor improvement capacitors, see NEMA Publication No. MG2, *Safety Standard for Construction and Guide for Selection, Installation and Use of Electric Motors and Generators.* Also see "Power Factor Corrections—Motor Circuit" under Article 460 of the National Electrical Code.

The level to which the power factor should be improved depends on the economic payback in terms of the electric utility power factor penalty requirements and the system energy saved due to lower losses. In addition the characteristic of the motor load must be considered. If the motor load is a cyclical load that varies from the rated load to a light load, the value of corrective kvar capacitance should not result in a leading power factor at light loads.

To avoid this possibility, it is recommended that the maximum value of the corrective kvar added be less than the motor no load kvar requirement by approximately 10 percent. Thus

Maximum capacitor kvar for
three-phase motors
$$= \frac{I_{NL} V \sqrt{3} \times 0.09}{1000}$$

$$= \frac{I_{NL} V}{642}$$

where
I_{NL} = motor no load line current
V = motor line voltage

For example, consider a 50-hp, 1750-rpm motor operating on a 230-V, three-phase, 60-Hz power system. Table 4.2 shows the motor performance without the power factor correction. Table 4.3 shows the combination motor-capacitor performance for various values of capacitor kvars up to a system power factor of 98.8 percent. Note the line loss reduction with the improved power factor. Based on 4000 hr/yr of operation at 5¢/kWh for electric energy, a correction to a 95 percent power factor will result in an electric energy saving of $52/yr.

The system performance at lighter motor loads is shown in Table 4.4. Note that at values of capacitors kvars 15 and larger the system power factor is leading. This is an unsatisfactory operating condition and should be avoided. Checking the maximum value of capacitance recommended based on the motor no load line current, we note that

$$\text{Maximum capacitor kvar} = \frac{35 \times 230}{642} = 12.54 \text{ kvar}$$

which confirms the change to a leading power factor as shown in Table 4.4. Figure 4.6 illustrates the change in the power factor at different loads and correction capacitor kvars.

Where to Locate Capacitors

The power factor correction capacitors should be connected as close as possible to the low-power-factor load. This is very often determined

Table 4.2 Induction Motor Performance Without a Power Factor Correction[a]

Load	Line amps	Eff	PF	kW input	kVA	kvar
Full load	119	0.915	0.862	40.8	47.4	24.1
¾ load	91	0.922	0.852	30.8	36.2	19.0
½ load	66	0.920	0.785	20.7	26.3	16.2
¼ load	45	0.887	0.598	10.8	17.9	14.3
No load	35	0	0.073	1.0	13.9	13.9

[a]50-hp, 1750-rpm, 230-V, three-phase, 60-Hz motor.

Table 4.3 Induction Motor and Capacitor Performance with a Power Factor Correction at Full Load[a]

Capacitor kvar	System line current	System kVA	System PF	Line loss reduction (%)
0	119	47.4	0.862	0
4	114	45.5	0.897	8
8	110	43.9	0.929	15
12	107	42.6	0.958	19
15	105	41.8	0.976	22
18	104	41.3	0.988	24

[a]50-hp, 1750-rpm, 230-V, three-phase, 60-Hz motor and capacitor.

Table 4.4 Induction Motor and Capacitor Performance with Power Factor Correction at Various Motor Loads[a]

Capacitor kvars	Motor load				
	Full load	¾ load	½ load	¼ load	No load
0	0.862	0.852	0.785	0.598	0.073
4	0.897	0.887	0.860	0.720	0.130
8	0.929	0.042	0.930	0.860	0.170
12	0.958	0.975	0.980	0.980	0.470
15	0.976	0.994	1.000	(0.99L)	(0.67L)
18	0.988	1.000	(0.99L)	(0.95L)	(0.24L)

[a]50-hp, 1750-rpm, 230-V, three-phase, 60-Hz system.

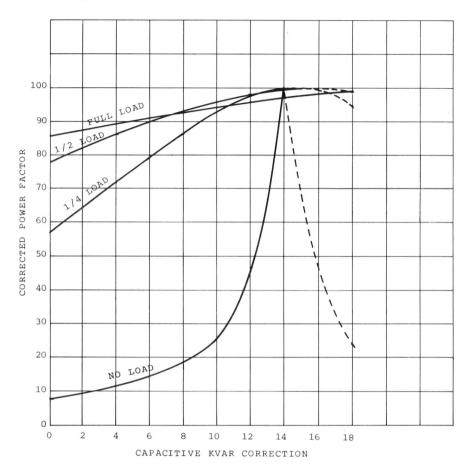

Fig. 4.6 Power factor at various loads and correction capacitors; the dashed line is a leading power factor.

by the nature and diversity of the load. Figure 4.7 illustrates typical points of installation of capacitors:

1. At the Motor Terminals. Connecting the power capacitors to the motor terminals and switching the capacitors with the motor load is a very effective method for correcting the power factor. The benefits of this type of installation are the following: No extra switches or protective devices are required, and line losses are reduced from the point of connection back to the power source. Corrective capacitance

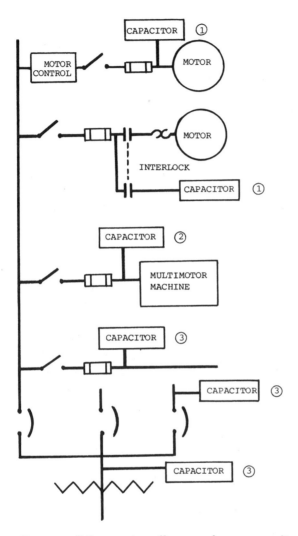

Fig. 4.7 Where to install power factor capacitors.

is supplied only when the motor is operating. In addition, the correction capacitors can be sized based on the motor nameplate information, as previously discussed.

If the capacitors are connected on the motor side of the overloads, it will be necessary to change the overloads to retain proper overload protection of the motor. A word of caution: With certain types of electric motor applications, this method of installation can result in damage to the capacitors or motor or both.

Never connect the capacitors directly to the motor when

Solid-state starters are used

Open transition starting is used

The motor is subject to repetitive switching, jogging, inching, or plugging

A multispeed motor is used

A reversing motor is used

There is a possibility that the load may drive the motor (such as a high-inertia load)

In all these cases, self-excitation voltages or peak transient currents can cause damage to the capacitor and motor. In these types of installations, the capacitors should be switched with a contactor interlocked with the motor starter.

2. At the Main Terminal for a Multimotor Machine.   In the case of a machine or system with multiple motors, it is common practice to correct the entire machine at the entry circuit to the machine. Depending on the loading and duty cycle of the motors, it may be desirable to switch the capacitors with a contactor interlocked with the motor starters. In this manner, the capacitors are connected only when the main motors of a multimotor system are operating.

3. At the Distribution Center or Branch Feeder.   The location of the capacitors at the distribution center or branch feeder is probably most practical when there is a diversity of small loads on the circuit that require power factor correction. However, again the capacitors should be located as close to the low-power-factor loads as possible in order to achieve the maximum benefit of the installation.

In all cases, regardless of the location of the capacitors, the installation should be in accordance with Article 460 of the National Electric Code.

4.4 POWER FACTOR MOTOR CONTROLLERS

In recent years, solid-state control devices have been developed which, when connected between a power source and an electric motor, maintain an approximately constant power factor on the motor side of the controller. These devices are generally called power factor controllers. Most of the units are made under a license of U.S. Patent 4,052,648 issued to F. J. Nola and assigned to NASA.

The controller varies the average voltage applied to the motor as a function of the motor load and thus decreases the motor losses at light load requirements.

Single-Phase Motors

For application to single-phase motors, the power factor controller consists of a triac, sensing and control circuits, and a firing circuit for the triac, as shown in Fig. 4.8. The power factor controller sensing circuit monitors the phase angle between the voltage and current and produces a signal proportional to the phase angle. This signal is compared to a reference signal which indicates the desired phase angle. This comparison produces an error signal which provides the timing for firing the triac or SCR and causes the phase angle to remain constant when the load changes. Typical motor voltage and current waveforms are shown in Figs. 4.9 and 4.10.

If the phase angle increases, the control circuit adjusts the triac firing angle to decrease the average voltage applied to the motor. Con-

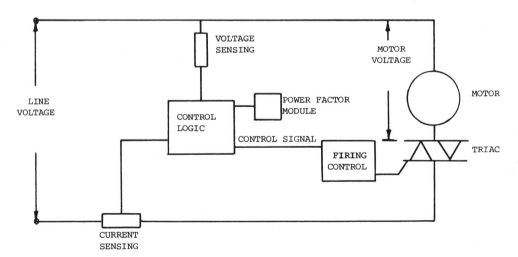

Fig. 4.8 Single-phase power factor controller block diagram.

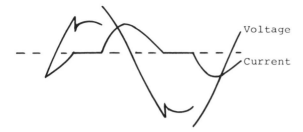

Fig. 4.9   Single-phase power factor controller with no load on the motor.

versely, if the phase angle decreases, the control circuit adjusts the firing angle of the triac to increase the average voltage applied to the motor.

The power factor of the motor is the cosine of the phase angle between the motor voltage and current. Therefore, with this control system, by maintaining the phase angle constant, the motor operates at an approximately constant power factor over the load range. The maximum power factor is the power factor of the motor at the rated load with the triac full on. The minimum power factor will be determined by the minimum voltage setting for no load operation. This voltage setting must be high enough to provide stable operation and prevent the motor from stalling on the sudden application of load. However, the lower the no load voltage, the higher the power savings at no load.

How are power savings achieved by decreasing the motor voltage at light loads? The motor losses can be grouped into three categories:

1. Constant losses such as friction and windage
2. Magnetic core losses which are some function of the applied voltage
3. I^2R losses which are a function of the square of the motor current, including rotor losses

For a given load condition, the net losses, and hence the motor power input, will decrease with a decrease in voltage as long as the magnetic core losses decrease more than the I^2R losses increase. In addition, there will be some increase in losses due to harmonics added to the motor input voltage by the triac switching and the losses in the controller.

In some instances, the increased harmonic content of the input voltage will result in increased motor noise.

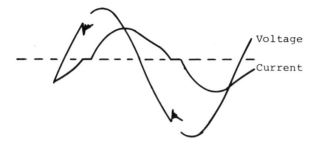

Voltage

Current

Fig. 4.10   Single-phase power factor controller with a full load on the motor.

The amount of power saved with a power factor controller depends on the duty cycle of the application. Typical power savings under various loads and duty cycles are shown in Figure 4.11. The power savings are shown as a percent of the full voltage input and as a function of the percent running times at full load versus running at a light load. To result in significant power savings, at least 50 percent of the running time should be at one-fourth load or less. Typical applications of this type may be drill presses and cutoff saws used in production processes.

Figure 4.9 shows an oscilloscope picture of the motor voltage and current at no load for a single motor controlled by a power factor controller.

Figure 4.10 shows an oscilloscope picture of the motor voltage and current of the same motor with load applied to the motor. Note the constant angle between the 0 crossing of the voltage and current in both cases.

Three-Phase Motors

More recently, the application of power factor motor controllers has been extended to apply to three-phase motors. In some cases, this has been accomplished by adding a power saver module to existing solid-state three-phase motor controllers. These solid-state controllers generally include other features such as current limit, timed acceleration, phase unbalance, undervoltage, and overload protection.

The power factor control function is accomplished by sensing the phase angle between the motor voltage and current. This signal is fed back and compared with a reference, and the difference is used to feed the input signal voltage to the six SCRs in the power module. The feedback voltage from the power factor sensing circuit will change the average voltage applied to the motor in accordance with the load on the motor. This will reduce both the motor current and voltage under light load conditions. The circuit is designed to react to load changes to prevent stalling of the motor on instantaneous load changes. Most of the controllers have provisions for setting the minimum no load voltage; this voltage is generally 65 percent of rated full voltage. Figure 4.12 is a typical block diagram for the three-phase controller.

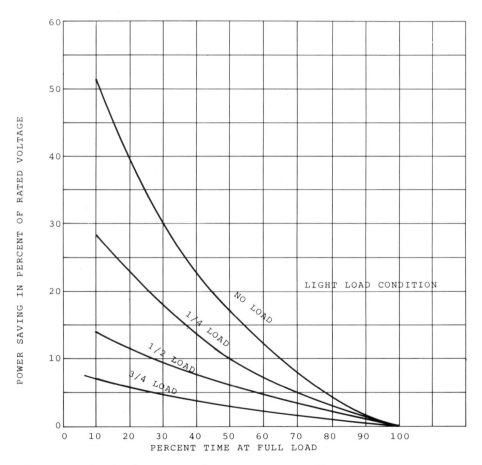

Fig. 4.11   Single-phase power factor motor controller power savings.

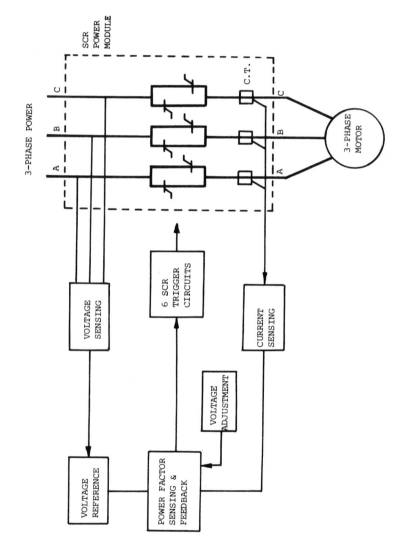

Fig. 4.12 Three-phase power factor motor controller block diagram.

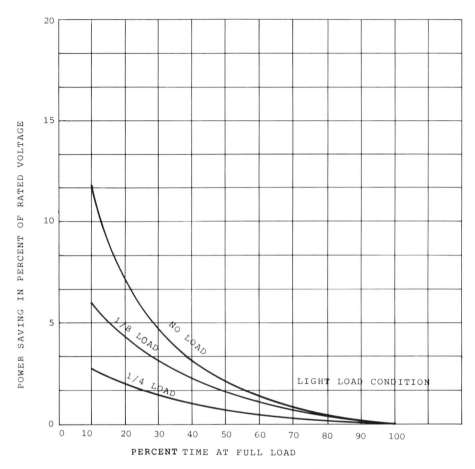

Fig. 4.13   Three-phase power factor motor controller power savings.

The three-phase power factor controllers have potential applications where the duty cycle for the motor is varying from light or no
load to full load as a step function. Examples of potential applications
are rip saws, conveyors, rock crushers, and centrifuges.

The potential power savings when a power factor controller is
applied to a three-phase motor is substantially lower than when one is
applied to a single-phase motor. Figure 4.13 illustrates the power savings when the controller is applied to a three-phase motor for various
duty cycles and loads. These curves will depend on the ratio of the no
load losses of the motor. However, it appears that the power factor

controller will only show significant power savings on those three-phase motor applications where the motor operates at no load or light loads over 75 percent of the operating time.

To apply a power factor controller properly, the load characteristics, motor characteristics, and load cycle must be known. In addition, one must determine how the controller-motor combination will respond to the load cycle. Only then can the potential power savings and economic payback analysis be made.

5

Application of Electric Motors

5.1 GENERAL DISCUSSION

The AC induction motor is used more than any other means to power industrial equipment. This is confirmed by the U.S. Department of Energy report on electric motors which states that 53 to 58 percent of the electric energy generated is consumed by electric motors (see Table 5.1). Because there are so many applications, it is impossible to develop a list or guide for all the applications of AC induction motors.

However, a guide for the selection of three-phase induction motors for many applications had been developed and is shown in Table 5.2. The table outlines the matching of the driven load requirements to the electric motor characteristics. In applying this guide it must be recognized that the purpose of the horsepower ratings and NEMA standards for electric motors is to define the motors' useful range of performance in a way most intelligible to the user. The rating of induction motors includes six major variables:

Supply voltage and frequency
Number of phases
Rated horsepower
Torque characteristics
Speed
Temperature rise

In addition, the basis of rating specifies the type of duty:

Table 5.1 Electric Motor Population and Energy Consumption, 1977

Horsepower	No. of motors (000s)	New sales, average for 1973–1977 (000s)	Annual electric energy consumption (billions of kWh)
1–5	54,583	3567	34
5.1–20	10,421	573	103
21–50	3,313	151	155
51–125	1,703	59	338
126 and over	1,004	35	573
Total	71,024	4385	1203

Source: U.S. Department of Energy Report DOE/CS-0147, 1980.

Continuous duty
Intermittent duty
Varying duty

It is desirable to use standard motors for as many different applications as possible. Consequently, general-purpose continuous rated motors should be used when

1. The peak momentary overloads do not exceed 75 percent of the breakdown torque
2. The root-mean-square (RMS) value of the motor losses over an extended period of time does not exceed the losses at the service factor rating
3. The duration of any overload does not raise the momentary peak temperature above a value safe for the motor's insulation system

Energy-Efficient Motors

The selection of an energy-efficient motor should be based on several additional factors:

1. Electric power-saving and life-cycle-cost comparison to standard motors.
2. Improved ability to perform under adverse conditions such as abnormal voltage. (See Secs. 5.3 and 5.4 for performance compar-

Table 5.2 Three-Phase Electric Motor Selection Chart

For this type of equipment	Requiring these torques		With these load characteristics	Type and description[a]
	Starting	Max. running		
Water supply pumps Industrial and chemical pumps Cooling towers Air-handling equipment Compressors Conveyors Process machinery Petroleum and chemical process equipment	100–150% of full-load torque	200–250% of full-load torque	Continuous operation, constant speed, high speed (over 720 rpm), easy starting; subject to short time overloads; good speed regulation	*Energy efficient:* NEMA design B, normal torques: normal starting current; can be used with variable-frequency/variable-voltage inverters; higher efficiency than standard design B motors
Centrifugal pumps Blowers and fans Drilling machines Grinders Lathes Compressors Conveyors	100–150% of full-load torque	200–250% of full-load torque	Variable load conditions, constant speed; subject to short time overloads; good speed regulation	*NEMA design B,* normal torques: normal starting current; can be used with variable-frequency/variable-voltage inverters

(continued)

Table 5.2 (*continued*)

For this type of equipment	Requiring these torques		With these load characteristics	Type and description[a]
	Starting	Max. running		
Reciprocating pumps Stokers Compressors Crushers Ball and rod mills	200–300% of full-load torque	Not more than full-load torque	High starting torque due to high inertia, back pressure, stand-still friction, or similar me-chanical conditions; torque requirements decrease during acceleration to full-load torque; not subject to severe over-loads; good speed regulation	NEMA *design C,* high torque: normal starting current; not recommended for use with variable-frequency inverters
Punch presses Cranes Hoists Press brakes Shears Oil well pumps Centrifugals	Up to 300% of full-load torque	200–300% of full-load torque; loss of speed dur-ing peak loads re-quired	Intermittent loads; may require frequent start, stop, and re-verse cycles; machine uses a flywheel to carry peak loads; poor speed regulation to smooth power peaks; may require acceleration of high-inertia load	NEMA *design D,* high torque: high slip; standard types have slip characteristics of 5–8% or 8–13% slip

Applications	Starting torque	Starting torque value	Characteristics	Type
Blowers Fans Machine tools Mixing machines Conveyors Pumps	Some require low torque; others require several times full-load torque	200% of full-load torque at each speed	Speed selection is desired, and two, three, or four fixed speeds are sufficient; starting torque can be low on blowers to high on conveyors; metal cutting machines are usually constant hp; Friction loads (conveyors) are usually constant torque; fluid or air loads (blowers) are variable torque	*Multispeed:* general normal torque on dominant winding or speed; consequent pole windings or separate windings for each speed; based on load requirement, can be constant horsepower, constant torque, variable torque
Crushers Conveyors Bending rolls Ball and rod mills Centrifugal blowers Pumps Printing presses Cranes and hoists Centrifugals	Can provide torque up to maximum torque at standstill	200–300% of full-load torque	Loads that require very high starting torque with low starting current; require speed adjustment over limited range (2 to 1); torque control during acceleration or controlled acceleration	*Wound rotor:* requires rotor control system to provide desired characteristic; control may be resistors or reactors or fixed-frequency inverter in the secondary (rotor) circuit; actual load speed depends on setting of rotor control

aSee Chap. 1 for a detailed description.

97

isons to standard motors. Note the superior performance of
energy-efficient motors under abnormal voltage conditions.)

3. Lower operating temperatures.
4. Noise level.
5. Ability to accelerate higher-inertia loads than standards motors.
6. Higher operating efficiencies at all load points. (Figure 5.1 illus-
    trates this comparison on a 20-hp, 1765-rpm, polyphase induction
    motor. Note that at all loads the energy-efficient motor presents
    the opportunity for energy savings.)

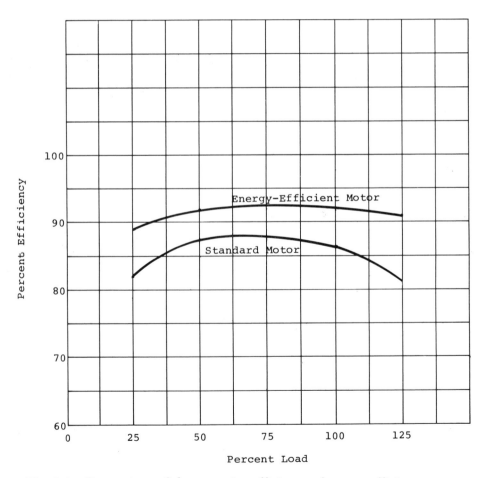

Fig. 5.1 Comparison of the operating efficiency of energy-efficient versus
standard polyphase induction motors at 20 hp and 1765 rpm.

In general, energy-efficient motors can be justified on a payback basis because of the annual saving of electric energy. This saving is a function of the hours of operation per year and kilowatt-energy reduction. For example, consider a 25-hp, 1800-rpm application with an average annual operating time of 4000 hr and a cost of electric power of 5¢/kWh:

Standard motor efficiency = 88 percent

$$\text{Power input} = \frac{25 \times 0.746}{0.88} = 21.19 \text{ kW}$$

Energy-efficient motor efficiency = 91.5 percent

$$\text{Power input} = \frac{25 \times 0.746}{0.915} = 20.38 \text{ kW}$$

Annual energy saving = $(21.19 - 20.38) \times 4000 = 3240 \text{ kWh}$

Annual power cost saving = $3240 \times 0.05 = \$162$

Typical list price standard dripproof motor = $616

Typical list price energy-efficient dripproof motor = $739

$$\text{Time to recover initial cost difference in years} = \frac{739 - 616}{162} = 9 \text{ months}$$

thus indicating a very favorable cost-benefit ratio for this application.

This method of cost-benefit analysis is approximate but is generally acceptable if the time to recover the initial investment is less than 3 yr. However, when it is desirable, more accurate methods can be used that consider the increasing cost of power, the required return on the investment, and the product useful life. The detailed procedures for making this type of economic analysis are described in Chap. 7.

In certain applications and duty cycles, energy-efficient motors cannot be justified on the basis of energy saved, for example,

1. Intermittent duty or special torque applications:
    Hoists and cranes
    Traction drives
    Punch presses
    Machine tools
    Oil field pumps
    Fire pumps
    Centrifugals
2. Types of loads:
    Multispeed

Frequent starts and stops
Very high-inertia loads
Low-speed motors (below 720 rpm)

Additional factors that should be considered in the selection and applications of electric motors are reviewed in the following sections of this chapter.

5.2   VARYING DUTY APPLICATIONS

In many applications the load imposed on the driving motor varies from no load to a peak load. When the motor load fluctuates, the temperature rise of the motor fluctuates. When there is a definite repeated load cycle, the motor size selection can be based on the RMS value of motor losses for the load cycle. However, normally the losses at each increment of the load cycle are not available to the user. Therefore, a good approximation for the motor size selection can be based on the RMS horsepower for the load cycle. The RMS horsepower is then defined as that equivalent steady-state horsepower which would result in the same temperature rise as that of the defined load cycle. When making the RMS calculation, it is assumed that when the motor is running the heat dissipation is 100 percent effective. However, when the motor is at standstill, the heat dissipation is severely reduced and is limited to dissipation by radiation and natural convection. This can be compensated for by using an effective cooling time at standstill of one-fourth of the total standstill time. *An important word of caution:* This method of selecting electric motors is not satisfactory for applications requiring frequent starting or plug reversing or systems with a high load inertia.

SAMPLE CALCULATION

$$
\begin{array}{lll}
\text{Duty cycle:} & \text{40 hp,} & \text{15 min} \\
& \text{20 hp,} & \text{20 min} \\
& \text{10 hp,} & \text{5 min} \\
& \text{Stop,} & \text{5 min} \\
\hline
& \text{Total cycle,} & \text{45 min}
\end{array}
$$

$$
\begin{aligned}
\text{hp}^2 \times t = (40)^2 \times 15 &= 24{,}000 \\
(20)^2 \times 20 &= 8{,}000
\end{aligned}
$$

$$(10)^2 \times 5 = \quad 500$$
$$0 \times 5 = \quad 0$$
$$\text{hp}^2 \times \text{t total} = 32{,}500$$

$$T_e = \text{effective cooling time} = 15 + 20 + 5 + \tfrac{1}{4} \times 5$$
$$= 41.25 \text{ min}$$

$$\text{rms hp} = \sqrt{\frac{\text{hp}^2 \times t}{T_e}} = \sqrt{\frac{32{,}500}{41.25}} = 28 \text{ hp}$$

From a thermal standpoint, a 30-hp standard motor would be satisfactory for this application.

Is the ratio of peak horsepower to nameplate (NP) horsepower satisfactory?

$$\frac{\text{Peak hp}}{\text{NP hp}} = \frac{40}{30} \times 100 = 133 \text{ percent}$$

Based on a limit of 150 percent for the ratio of peak horsepower to motor nameplate horsepower, the 30-hp motor could be satisfactory for this load.

Consider a slightly different cycle:

40 hp, 10 min
20 hp, 25 min
10 hp, 10 min

Total cycle, 45 min

$$(40)^2 \times 10 = 16{,}000$$
$$(20)^2 \times 25 = 10{,}000$$
$$(10)^2 \times 10 = 1{,}000$$

$$\text{hp}^2 \times \text{total} = 27{,}000$$

Effective cooling time $= 10 + 25 + 10 = 45$ min

$$\text{rms hp} = \sqrt{\frac{27{,}500}{45}} = 24.5 \text{ hp}$$

From a thermal standpoint, a standard 25-hp motor would be satisfactory.

However,

$$\frac{\text{Peak hp}}{\text{NP hp}} = \frac{40}{25} \times 100 = 160 \text{ percent}$$

Based on a limit of 150 percent for this ratio, the use of a 25-hp motor is not considered satisfactory.

5.3   VOLTAGE VARIATION

NEMA Standard MG1 recognizes the effect of voltage and frequency variation on electric motor performance. The standard recommends that the voltage deviation from the motor rated voltage not exceed ± 10 percent at the rated frequency. A certain degree of confusion may exist in regard to the rated motor voltage, since the rated motor voltage and the system voltage are different. The rated motor voltage has been selected to match the utilization voltage available at the motor terminals. This voltage allows for the voltage drop in the power distribution system and for voltage variation as the system load changes.

The basis of the NEMA standard rated motor voltages for three-phase, 60-Hz induction motors is as follows:

System voltage	Rated motor voltage
208	200
240	230
480	460
600	575

For single-phase, 60-Hz induction motors, the basis for standard rated motor voltages is as follows:

System voltage	Rated motor voltage
120	115
240	230

Polyphase electric motors are designed to operate most effectively at their nameplate rated voltage. Most motors will operate satisfactorily over ±10 percent voltage variation, but deviations from the nominal motor-design voltage can have marked effects on the motor performance. Table 5.3 indicates the type of changes in performance to expect with variation in the motor terminal voltage. The table shows the effect on the efficiency and power factor in standard NEMA design B motors and also in energy-efficient motors. It is important to note that the efficiency and the PF of energy-efficient motors are not as sensitive to voltage variations as standard motors.

In recent years, the trend in some areas is to decrease system voltage to reduce the system load. In some cases, this reduction has been as low as 85 percent of the nominal voltage. For most electric motor loads, this increases rather than decreases the electric motor input and increases the full-load temperature rise. Also, the locked-rotor torque is severely reduced such that hard-to-start loads may not start at the 85 percent voltage level. Figures 5.2 to 5.4 illustrate the effect of reduced voltage on selected horsepower ratings of both standard motors and energy-efficient motors.

5.4   VOLTAGE UNBALANCE

Voltage unbalance can be more detrimental to motor performance and motor life than voltage variation. When the line voltages applied to a polyphase induction motor are not equal in magnitude and phase angle, unbalanced currents in the stator windings will result. A small percentage voltage unbalance will produce a much larger percentage current unbalance.

Some of the causes of voltage unbalance are the following:

1.   An open circuit in the primary distribution system.
2.   A combination of single-phase and three-phase loads on the same distribution system, with the single-phase loads unequally distributed.
3.   An open wye-delta system
     a. Variation in ground supply impedance: An increase in primary ground impedance increases the voltage and current balances. Maximum unbalance occurs with overloaded transformers, and the large single-phase load is in the lagging phase. The motor serves to balance the system voltage better when the motor is loaded than when it is unloaded.

Table 5.3 Effect of Voltage Variation on Polyphase Induction Motor Performance

Operating characteristic	Effect of voltage change		
	90% voltage	110% voltage	120% voltage
Starting and maximum running torque	Decrease 19%	Increase 21%	Increase 44%
Synchronous speed	No change	No change	No change
Percent slip	Increase 23%	Decrease 17%	Decrease 30%
Full-load speed	Decrease 1½%	Increase 1%	Increase 1½%
Starting current	Decrease 10–12%	Increase 10–12%	Increase 25%
Magnetic noise, any load	Decrease slightly	Increase slightly	Noticeable increase
Standard NEMA design B motors Efficiency			
Full load	Increase ½–1%	Decrease 1–4%	Decrease 7–10%
3/4 load	Increase 1–2%	Decrease 2–5%	Decrease 6–12%
1/2 load	Increase 2–4%	Decrease 4–7%	Decrease 14–18%

Power factor			
Full load	Increase 8–10%	Decrease 10–15%	Decrease 10–30%
3/4 load	Increase 10–12%	Decrease 10–15%	Decrease 10–30%
1/2 load	Increase 10–15%	Decrease 10–15%	Decrease 15–40%
Full-load current	Increase 1–5%	Increase 2–11%	Increase 15–35%
Temperature rise at full load	Increase 6–12%	Increase 4–23%	Increase 30–80%
Energy-efficient NEMA B motors			
Efficiency			
Full load	Decrease 1–2%	Increase ½–1%	Small increase
3/4 load	Minimal change	Minimal change	Decrease ½–2%
1/2 load	Increase 1–2%	Decrease 1–2%	Decrease 7–20%
Power factor			
Full load	Increase 1%	Decrease 3%	Decrease 5–15%
3/4 load	Increase 2–3%	Decrease 4%	Decrease 10–30%
1/2 load	Increase 4–5%	Decrease 5–6%	Decrease 10–30%
Full-load current	Increase 11%	Decrease 7%	Decrease 11%
Temperature rise at full load	Increase 23%	Decrease 14%	Decrease 21%

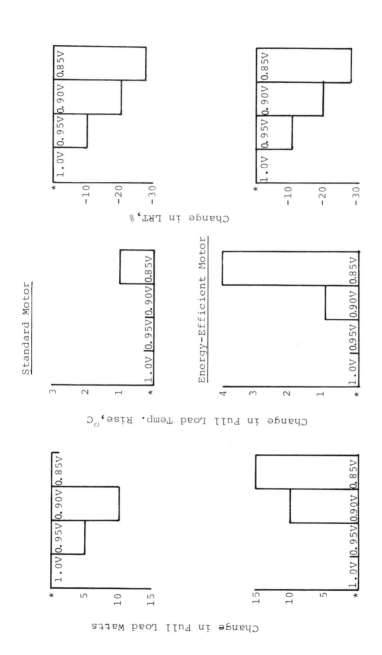

*Rating Point: Full-Load Watts, Full-Load Temperature Rise, and LRT at 100% Voltage

Fig. 5.2 Comparison of the effects of low voltage on the performance of three-phase standard and energy-efficient motors at 1 hp and 1750 rpm.

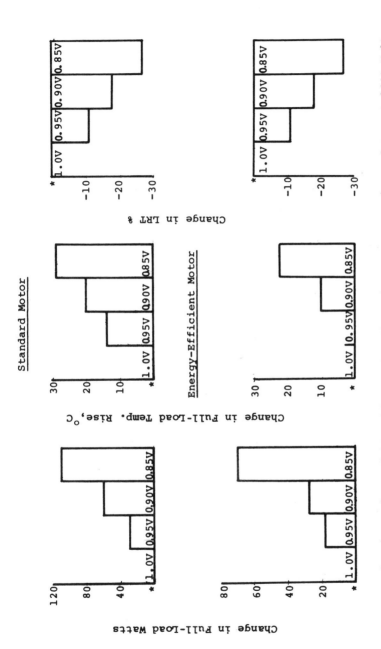

*Rating Point: Full-Load Watts, Full-Load Temperature Rise, and LRT at 100% Voltage

Fig. 5.3 Comparison of the effects of low voltage on the performance of three-phase standard and energy-efficient motors at 5 hp and 1750 rpm.

107

108

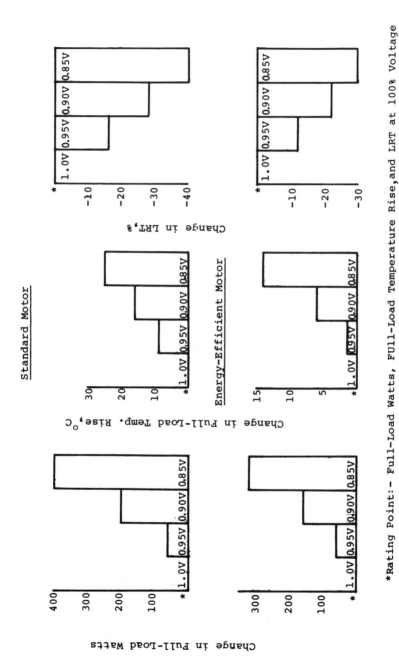

Standard Motor

Energy-Efficient Motor

*Rating Point:- Full-Load Watts, Full-Load Temperature Rise, and LRT at 100% Voltage

Fig. 5.4 Comparison of the effects of low voltage on the performance of three-phase standard and energy-efficient motors at 25 hp and 1750 rpm.

   b. Transformer loading varied 50 to 150 percent: The greatest unbalance occurs when a smaller transformer is lightly loaded and a larger transformer is overloaded. If a single-phase load varies over a large range, it is better to supply this phase with the larger transformer on the leading phase.

   c. Impedance of lines to the single-phase loads: The voltage and current unbalance ratios increase with the line impedances. Again, the unbalance ratios decrease as the motor is loaded more heavily.

   d. Impedance of the supply line to the motor: The voltage and current unbalance ratios decrease with an increase of the line impedance to the motor. However, this results in lower voltage at the motor and decreased motor torque and speed.

   e. Other parameters: Variations in the magnitude of transformer impedances, the power factor of single-phase loads, and primary line impedances have minor effects (not more than 3 percent) on the phase currents and unbalance ratios.

4. An open delta-delta system: When the two transformers are supplied by three-phase conductors, the only difference is in the lack of neutral impedance. Therefore, under usual conditions, the open delta-delta configuration will show superior performance to the open wye-delta configuration. However, when there are unequal line impedances or unusually long supply lines, there are additional observations.

   a. There are mixed effects with variation of the lines supplying the single-phase loads.

   b. An increase in the common primary supply line impedance results in increased voltage and current unbalances.

The unbalanced line voltages introduce negative sequence voltages in the polyphase motor. This negative sequence voltage produces an air gap flux rotating oppositely to the rotor, thus producing high currents in the motor. A small negative sequence voltage can produce motor currents considerably in excess of those present under balanced voltage conditions.

NEMA Standard MG1 defines the percent voltage unbalance as follows:

$$\text{Percent voltage unbalance} = 100 \times \frac{\text{maximum voltage deviation from average voltage}}{\text{average voltage}}$$

These unbalanced voltages will result in unbalanced currents on the order of 6 to 10 times the voltage unbalance. Consequently, the temperature rise of the motor operating at a particular load and voltage unbalance will be greater than for the motor operating under the same conditions with balanced voltages. In addition, the large unbalance of the motor currents will result in nonuniform temperatures in the motor windings. An example of the effect of unbalanced voltages on performance is illustrated in Table 5.4. for a 5-hp motor.

Voltages should be evenly balanced as closely as possible. *Operation of a motor above 5 percent voltage unbalance is not recommended.* Even at 5 percent voltage unbalance, motor current unbalance on the order of 40 percent can exist.

In recognizing the detrimental effect of unbalanced line voltage on electric motor performance, NEMA Standard MG1 recommends derating motors that are applied to unbalanced systems, in accordance with Fig. 5.5 (NEMA MG1-14.34 chart):

> When the derating factor is applied, the selection and setting of the overload device should take into account the combination of the derating factor applied to the motor and the increase in current resulting from the unbalanced voltages. This is a complex problem involving the variation in motor current as a function of load and voltage unbalance in addition to the characteristics of the overload device relative to $I_{maximum}$ or $I_{average}$. In the absence of specific information it is recommended that overload devices be selected and/or adjusted at the minimum value that does not result in tripping

Table 5.4   Effect of Voltage Unbalance on Motor Performance[a]

Characteristic	Performance		
Average voltage	230	230	230
Percent unbalanced voltage	0.3	2.3	5.4
Percent unbalanced current	0.4	17.7	40.0
Increased temperature rise, °C	0	30	40

[a]5-hp, 1725-rpm, 230-V, three-phase, 60-Hz motor.

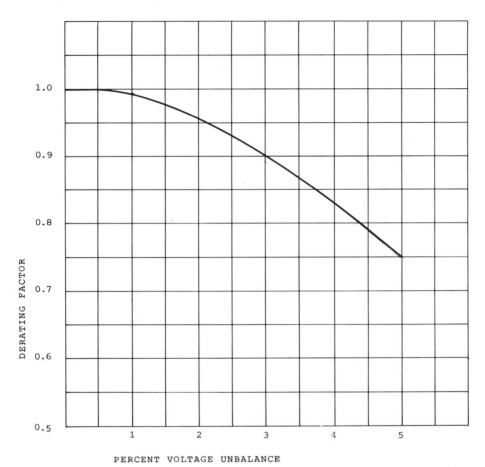

Fig. 5.5 Derating factor for unbalanced voltages on polyphase induction motors. (Reprinted by permission from NEMA Standards Publication No. MG1-1978, *Motors and Generators*, copyright 1978 by the National Electrical Manufacturers Association.)

for the derating factor and voltage unbalance that applies. When unbalanced voltages are anticipated, it is recommended that the overload devices be selected so as to be responsive to $I_{maximum}$ in preference to overload devices responsive to $I_{average}$.

The order of magnitude of the current unbalance is influenced by not only the system voltage unbalance but the system impedance, the nature of the loads causing the unbalance, and the operating load on the motor. Figure 5.6 indicates the range of unbalanced currents for various motor load conditions and system voltage unbalance.

The effect on other electric motor characteristics can be summarized as follows:

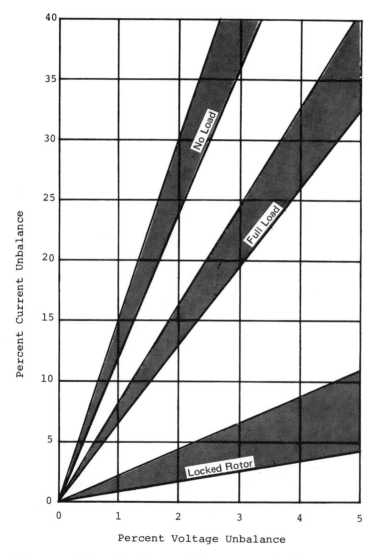

Fig. 5.6 Effect of voltage unbalance on polyphase induction motor currents.

Torques. The locked-rotor and breakdown torques are decreased. If the voltage unbalance should be extremely severe, the torques might not be adequate for the application.

Full-Load Speed. The full-load speed is reduced slightly.

Locked-Rotor Current. The locked-rotor current will be unbalanced to the same degree that the voltages are unbalanced, but the locked-rotor kVA will increase only slightly.

Noise and Vibration. The unbalanced voltages can cause an increase in noise and vibration. Vibration can be particularly severe on 3600-rpm motors.

5.5 OVERMOTORING

In many instances the practice has been to overmotor an application, i.e., to select a higher-horsepower motor than necessary. The disadvantages of this practice are the following:

Lower efficiency
Lower power factor
Higher motor cost
Higher controller cost
Higher installation costs

One example of overmotoring is illustrated by the case of the varying duty applications discussed in Sec. 5.2. Consider the comparisons of the 40-hp motor that could have been selected based on the peak load versus the 30-hp motor that can be selected on the basis of the duty cycle:

1. Motor cost: list price of standard open 1800-rpm dripproof motor:
    30 hp = $722
    40 hp = $908
2. Control Cost: NEMA-1 general-purpose motor, 240 V, starter:
    30 hp, size 3 = $230
    40 hp, size 4 = $526

This results in a cost difference of $482 or 51 percent.

Figure 5.7 shows the difference in the input watts and Fig. 5.8 the difference in the input kVA for 30- and 40-hp motors operating at the same output. At loads above 36 hp, the input is more favorable for the 40-hp motor. However, at loads below 36 hp, the kW and kVA inputs are lower with the 30-hp motor.

In general, the larger the difference between the actual load and the motor rating, the higher the input requirements for the same load.

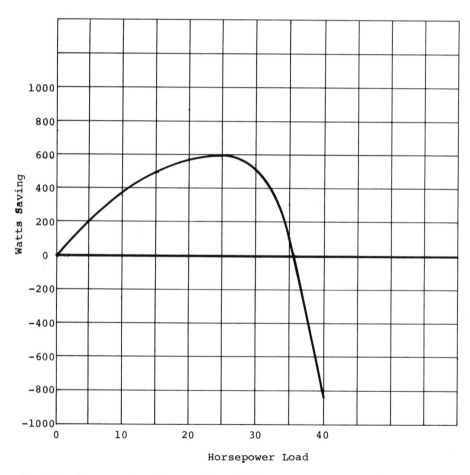

Fig. 5.7 Power savings in watts for a 30-hp motor versus a 40-hp motor at the same load.

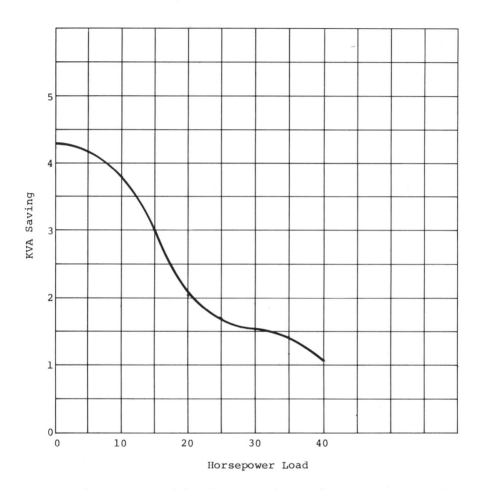

Fig. 5.8 kVA savings in kilovolt-amperes for a 30-hp motor versus a 40-hp motor at the same load.

6

Induction Motors and Adjustable-Speed Drive Systems

6.1 ENERGY CONSERVATION

The potential energy conservation possible in any system can best be determined by examining each element of the system and its contribution to the losses and inefficiency of the system. Every device that does any work or causes a change in state of a material has energy losses.

Thus typical losses include the following:

1. Electrical transmission losses from the metering point to the system (this is where the electric power consumption is measured and the power bill determined)
2. Conversion losses in any power conditioning equipment (this includes variable-frequency inverters and the effect of the inverter output on the motor efficiency)
3. Electric motor losses to convert electric power to mechanical power
4. Mechanical losses in devices such as gears, belts, and clutches to change the output speed of the motor
5. Losses in the driven unit such as a pump or fan or any other device that performs work on material
6. Transmission losses such as friction losses to move material from one location to another
7. Losses caused by throttling or other means to control material flow by absorbing or bypassing excess output

116

Each element in a particular system has an efficiency which can be defined as

$$E = \frac{\text{output power}}{\text{input power}}$$

or

Losses = input power − output power

The overall efficiency of the system is the product of the efficiencies of all elements of the system; thus,

$$E_{overall} = E_1 \times E_2 \times E_3 \times E_4 \times E_5$$

Therefore, the proper selection of each element can contribute to electric energy conservation.

This can be illustrated by an example of a constant-speed pumping system. The pumping system is to move water from one location to another at 1000 gpm with a static head of 100 ft. The friction head is 30 ft with a 4-in.-diameter supply pipe. What is the energy saving using a 5-in.-diameter supply pipe and an energy-efficient motor? Table 6.1 shows a summary of the calculations for the two systems. The net result is an annual savings of 28,520 kWh or 19.7 percent of the input. Note that the savings were achieved by improved performance in several elements: Lower motor losses due to improved efficiency and lower horsepower required, lower pump losses due to increased efficiency and lower horsepower required, and lower pipe friction losses with improved hydraulic efficiency. The overall system efficiencies are as follows:

4-in. system efficiency = 0.769 × 0.77 × 0.90 × 0.978 = 0.521
5-in. system efficiency = 0.909 × 0.79 × 0.92 × 0.982 = 0.649

The conclusion is that the complete system needs to be considered to obtain the most energy-efficient installation. One aspect not to be overlooked is that the losses in the system are dissipated as heat at each device such as the motors, pumps, compressors, etc. Therefore, if the devices are in a conditioned environment, the effect of the losses or change in the losses on the conditioning system must also be considered.

In many installations additional energy savings can be achieved by combining the fixed-speed induction motor with some method of

Table 6.1 Summary Calculations for Example of Pump Installation

	4-in. supply pipe	5-in. supply pipe
Static head, ft	100	100
Friction head, ft	30	10
Total head, ft	130	110
Output hydraulic hp	25.25	25.25
Input hydraulic hp	37.83	27.78
Hydraulic efficiency	0.769	0.909
Pump efficiency	0.77	0.79
Pump input, hp	42.64	35.16
Motor standard hp	50	40
Motor efficiency at operating load	0.90	0.92
Motor input, hp	47.38	38.22
Transmission efficiency	0.978	0.982
System efficiency	0.521	0.649
System input, hp	48.46	38.91
System input, kW	36.15	29.02
Energy saving, kW	—	7.13
Annual saving, kWh	—	28,520
Percent savings	—	19.7

varying the output speed of the unit. This is particularly true in any application where output is a fluid flow that must vary in response to some other variable. A similar opportunity exists if output pressure must be controlled with a varying flow or varying input pressure.

Many fluid processes (including air processes) involve pumping the fluid to a high pressure and controlling flow and pressure to the required levels by throttling or bypassing. These throttling and bypass methods of control are inherently inefficient.

Centrifugal pumps, fans, and blowers have characteristics in accordance with the laws of fan performance. These laws state the following:

Flow varies directly with speed.
Pressure varies as the square of the speed.
Power varies as the cube of the speed.

These types of applications lend themselves to conversion from throttled constant-speed systems to adjustable-speed systems and offer a large potential for energy savings.

6.2 ADJUSTABLE-SPEED SYSTEMS

There are many types of adjustable-speed systems available. Some of the more popular types of adjustable-speed drives are the following: multispeed motors, adjustable-speed pulley systems, mechanical adjustable-speed systems, eddy current adjustable-speed drives, fluid drives, DC adjustable-speed systems, AC variable-frequency systems, and wound rotor motors.

The selection of the most effective system for a specific application depends on a number of factors:

Life-cycle cost
First cost
Duty cycle and horsepower range
Energy consumption
Control features required
Size
Performance
Reliability
Maintenance

To assist in the selection of an adjustable-speed drive system, let us examine the characteristics of the more popular ones. The DC adjustable-speed systems have been specifically excluded from this section since their characteristics and application technology is well known to those who apply them.

Multispeed Motors

As discussed earlier, multispeed motors can be obtained with the following output characteristics:

Constant horsepower
Constant torque
Variable torque

However, in conventional multispeed motors, only a limited variety of speed combinations is available. One-winding, two-speed motors are available with 2 to 1 speed combinations such as

1750 rpm/850 rpm
1150 rpm/575 rpm

Two-winding, two-speed motors are available with speed combinations other than 2 to 1. Typical speed combinations are

1750 rpm/1150 rpm
1750 rpm/850 rpm
1750 rpm/575 rpm
1150 rpm/850 rpm

Thus there are more combinations of speed ratios available in the two-winding, two-speed motors.

In addition, two-winding, four-speed motors are also available. A typical speed combination is 1750 rpm/1150 rpm/850 rpm/575 rpm.

Since the power requirements for many fans and centrifugal pumps are a cube of the speed, the variable-torque multispeed motor can be used for two-step speed control, i.e., a high-speed and a low-speed operation. With a one-winding, two-speed motor the output of the fan or pump on low-speed will be 50 percent of the output on high speed, and the horsepower required will be 12.5 percent of high speed. Figure 1.7 shows a fan load curve superimposed on the speed-torque curves for a variable-torque multispeed motor. In the case of a two-winding, two-speed motor with a combination of 1750 rpm/1150 rpm, the output of the fan or pump on low speed will be 67 percent of the output on high speed, and the horsepower required will be 30 percent of high speed. This is illustrated by Fig. 1.8, which shows a fan load curve superimposed on the motor speed-torque curves. Recognizing the speed limitations of this type of drive, it is an economical and reliable method to obtain incremental flow control.

Adjustable-Speed Pulley Systems

An adjustable-speed pulley system consists of the electric motor mounted on a special base, an adjustable-speed sheave on the motor shaft, and a fixed-diameter sheave on the load shaft connected by a V belt. Figure 6.1 shows the construction of one of the variable-speed sheaves, and Fig. 6.2 shows the special base required for the drive motor. The V belt in the spring-loaded motor sheave changes its diametric position, resulting in a change in the ratio of the effective pitch diameters of the two sheaves and a change in speed. This type of drive has limited capacity up to approximately 125 hp and a limited speed range of approximately 2 or 3 to 1. However, the efficiency over a limited range is on the order of 80 to 90 percent. Its major disadvantage is

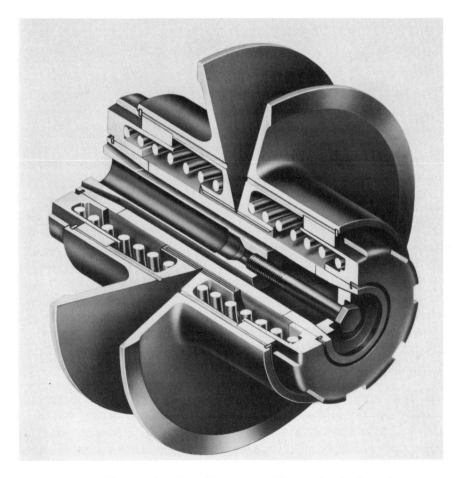

Fig. 6.1 Variable-speed pulley. (Courtesy of T. B. Wood's Sons Company, Chambersburg, Pa.)

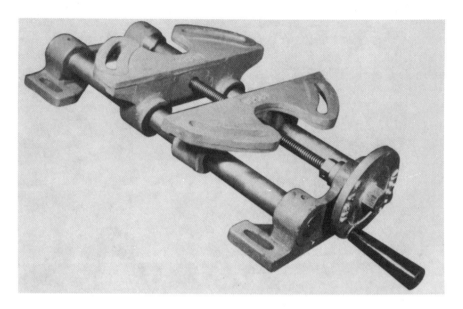

Fig. 6.2 MBA motor base. (Courtesy of T. B. Wood's Sons Company, Chambersburg, Pa.)

that the speed must be changed manually and does not lend itself to automatic or remote control. However, for applications that require only occasional adjustment in output, this system may be adequate and still provide energy savings at the lower speed settings, for example, an air-handling system that requires output adjustment only for summer and winter operation. Figure 6.3 shows a typical installation of this type of drive.

Mechanical Adjustable-Speed Systems

The broad group of mechanical adjustable-speed drives includes the more common stepless mechanical adjustable-speed drives that provide an infinite number of speed ratios within a nominal speed range. These types of drives include packaged belt and chain drives, friction drives, and traction drives. These drive systems are usually driven by a constant-speed induction motor and convert this constant-speed input into a stepless variable-speed output. A typical group of packaged adjustable-speed belt-drive systems is shown in Fig. 6.4. In the case of the

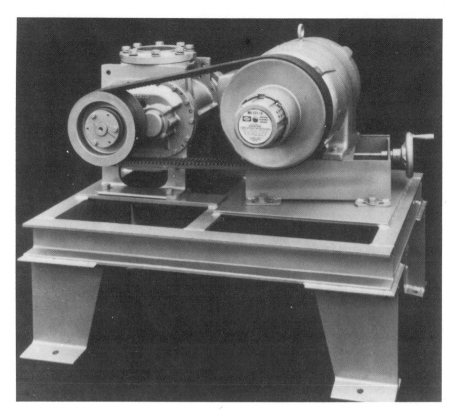

Fig. 6.3 Installation of variable-speed sheave. (Courtesy of T. B. Wood's Sons Company, Chambersburg, Pa.)

Fig. 6.4 U.S. Electrical Motors varidrive units. (Courtesy of U.S. Electrical Motors, division of Emerson Electric Co., Milford, Conn.)

belt-drive systems, the basis of rating is generally constant torque (variable horsepower) at speed ratios below 1 to 1 and constant horsepower (variable torque) at speed ratios above 1 to 1. Figure 6.5 illustrates this basis of rating. The efficiency of these systems at various loads and speeds is shown in Figs. 6.6 and 6.7 for representative ratings of the U.S. varidrive line of packaged mechanical belt drives.

Most of these types of drives have limited horsepower and speed ranges. Therefore, the selection of the drive systems should be based on the duty cycle of the load and the characteristics of the drive under

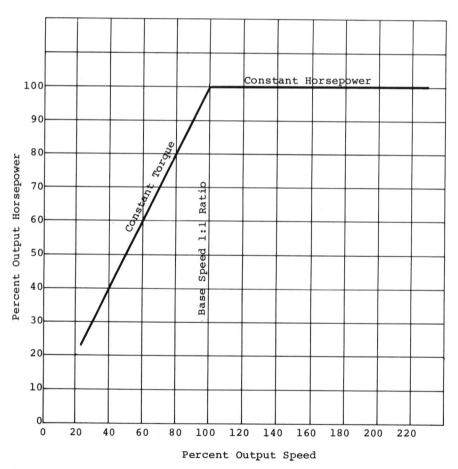

Fig. 6.5 Horsepower output versus output speed of a mechanical adjustable-speed drive system. (Courtesy of U.S. Electrical Motors, division of Emerson Electric Co., Milford, Conn.)

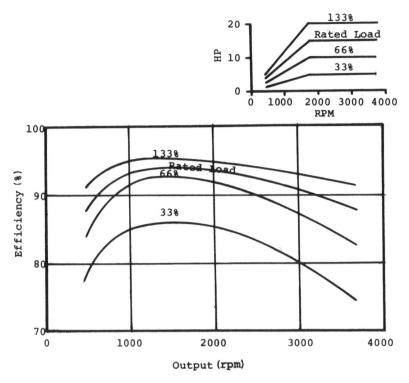

Fig. 6.6 Varidrive performance curves, typical data, for a 15-hp, four-pole motor. (Courtesy of U.S. Electrical Motors, division of Emerson Electric Co., Milford, Conn.)

consideration including speed range, horsepower, torque characteristics, and efficiency over the duty cycle. When properly applied, many of these drives have good efficiencies over their operating range. It is recommended that several types of systems be compared to determine the most suitable and effective life-cycle cost system. The requirements and type of remote control must also be a factor in the selection of the drive system.

Figures 6.8, 6.9, and 6.10 illustrate the types of process controls that are available on packaged belt drives.

Eddy Current Adjustable-Speed Drives

An eddy current adjustable-speed drive system consists of a constant-speed drive unit (usually an induction motor), an eddy current cou-

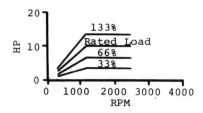

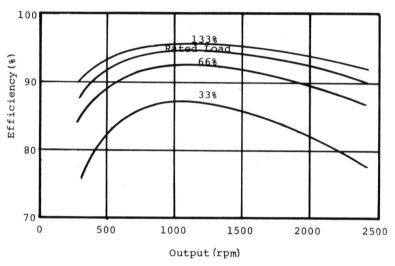

Fig. 6.7 Varidrive performance curves, typical data, for a 10-hp, six-pole motor. (Courtesy of U.S. Electrical Motors, division of Emerson Electric Co., Milford, Conn.)

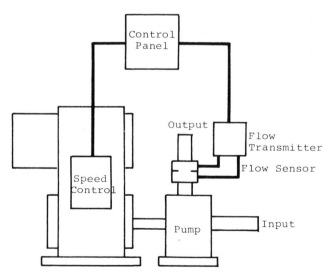

Fig. 6.8 Output flow control of a mechanical adjustable-speed system.

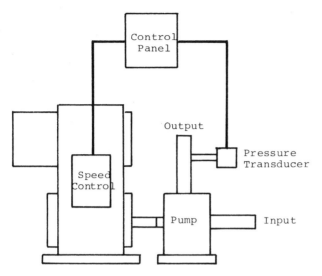

Fig. 6.9 Line pressure control of a mechanical adjustable-speed system.

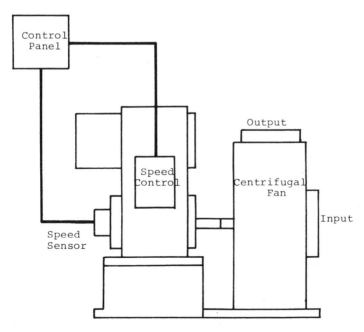

Fig. 6.10 Speed control of a mechanical adjustable-speed system.

pling, and a load speed control system. The eddy current coupling can be either a self-contained unit, as shown in Fig. 6.11, or a combined unit with the drive motor and eddy current coupling in a single housing, as shown in Fig. 6.12. The eddy current coupling consists of input and output members which are mechanically independent, with the output member revolving freely within the input member. The output member is the magnetic member and has a field winding which is excited by a DC current. The application of the field current creates a flux across the air gap between the two members which induces eddy currents in the input member. The net result is a torque available at the output shaft. A change in field current will change the torque output. Therefore, by adjusting the field current, the output can be adjusted to match the output load speed requirements. The power flow for this type of system is shown in Fig. 6.13.

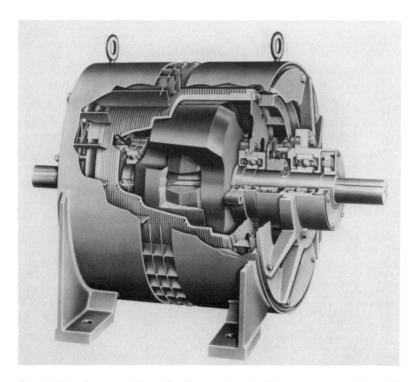

Fig. 6.11   Cross-section of self-contained eddy current coupling. (Courtesy of Electric Machinery, Power Systems Group, McGraw-Edison Company, Minneapolis.)

Fig. 6.12 Integral eddy current coupling and electric motor. (Courtesy of Electric Machinery, Power Systems Group, McGraw-Edison Company, Minneapolis.)

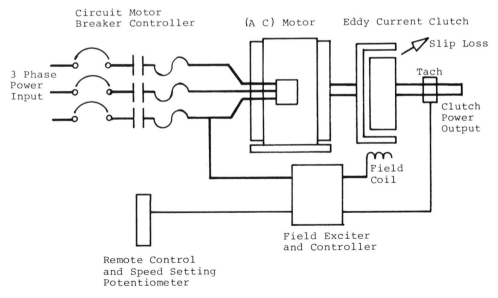

Fig. 6.13 Power flow for eddy current drive system.

The degree of coupling or slip between the two members is determined by the load and level of excitation. The slipping action (i.e., difference in speed) is the source of the major power loss and inefficiency of the eddy current coupling. This slip loss is the product of the slip rpm, which is the difference in speed between the input and output members and the transmitted torque. This relationship may be expressed as follows:

$$\text{Load (output) hp} = \frac{\text{rpm}_2 T_L}{5250}$$

$$\text{Motor (input) hp} = \frac{\text{rpm}_1 T_L}{5250}$$

$$\text{Slip loss} = \text{motor hp} - \text{load hp}$$

$$= \frac{T_L(\text{rpm}_1 - \text{rpm}_2)}{5250}$$

where

rpm_1 = coupling input speed (motor)

rpm_2 = coupling output speed (load)

T_L = load torque, ft-lb

The efficiency of an eddy current coupling can never be greater than the numerical percentage of the output speed. However, in addition to the slip losses, the friction and windage losses and excitation losses of the coupling must also be included in the efficiency determination. The friction and windage loss is 1 to 3 percent of the rated input horsepower and can be considered constant over the speed range. The excitation loss is less than 0.5 percent of the input horsepower and decreases with reduction in speed. Typical speed-torque curves for various levels of excitation are shown in Fig. 6.14. Superimposed on the curves is the torque curve for a fan or centrifugal pump in a friction–only system. Figure 6.15 illustrates the relationship of fan horsepower, motor horsepower, losses, and efficiency for a specific fan application. Centrifugal pumps have similar variable-torque characteristics but usually operate against some static head. This causes the pump torque to follow a closed discharge characteristic until sufficient speed is reached to cause the resultant discharge head to equal or exceed the static head. The pump load then follows a curve different from the full-load, full-speed condition. This type of pump load is illustrated in Fig.

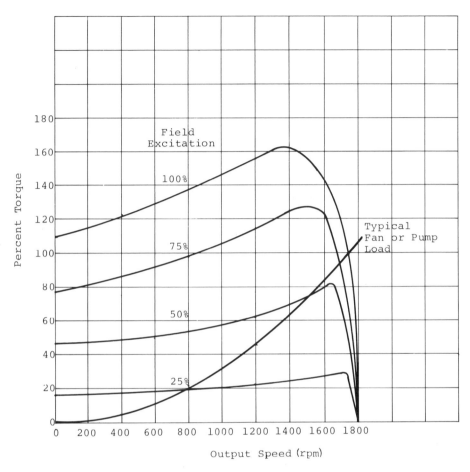

Fig. 6.14 Typical eddy current drive speed-torque curves. (Courtesy of Electric Machinery, Power Systems Group, McGraw-Edison Company, Minneapolis.)

6.16. Curve 0AC indicates the pump load horsepower; at point A the static head is overcome, and the pump load rises along curve AC as the speed is increased. However, under the conditions indicated at speeds below point A, the static head is not exceeded, and there is no flow.

Since the eddy current coupling has no inherent speed regulation, it is necessary that the coupling include a tachometer generator which rotates at the coupling output speed. The tachometer-generator out-

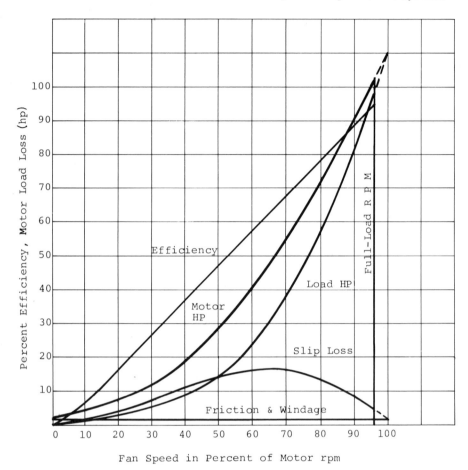

Fig. 6.15 Magnetic drive characteristics for fan load. (Courtesy of Electric Machinery, Power Systems Group, McGraw-Edison Company, Minneapolis.)

put signal is fed into a speed control loop in the excitation system to provide close output speed regulation. The speed regulation is usually ±1 percent but with closed-loop control may be as close as ±0.1 percent. In addition to speed, the eddy current control system can be used with any type of actuating device or transducer which can provide a mechanical translation or an electrical signal. Actuating devices include liquid-level control, pressure control, temperature control, and flow control. The block diagram for a pressure control system is shown in Fig. 6.17. In this instance, the control action requires the pressure controller to demand a pump speed which will provide a feedback voltage from the pressure transducer to match a pressure reference voltage. When these voltages are unequal, the error voltage results in a corrective speed change.

Figure 6.18 shows an eddy current system driving a vertical pump installation.

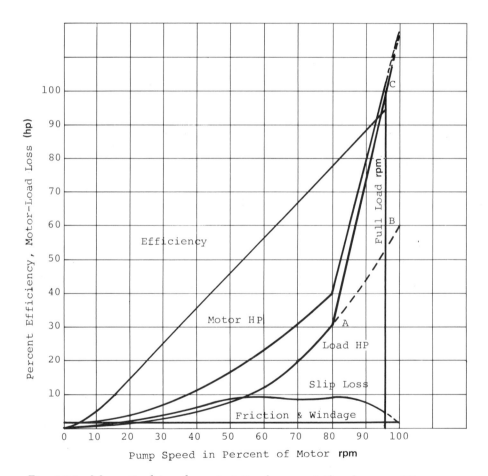

Fig. 6.16 Magnetic drive characteristics for a centrifugal pump. (Courtesy of Electric Machinery, Power Systems Group, McGraw-Edison Company, Minneapolis.)

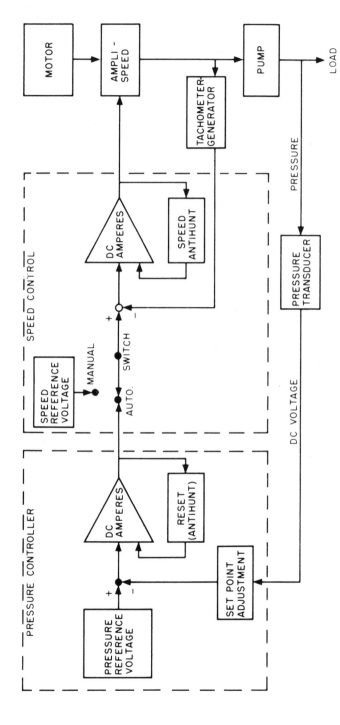

Fig. 6.17 Block diagram of the pressure control system for eddy current drive. (Courtesy of Electric Machinery, Power Systems Group, McGraw-Edison Company, Minneapolis.)

Fig. 6.18 Typical installation of a vertical induction motor and an eddy current drive on a pump. (Courtesy of Electric Machinery, Power Systems Group, McGraw-Edison Company, Minneapolis.)

Fluid Drives

Fluid drives can be described as any device utilizing a fluid to transmit power. The fluid usually used is a natural or synthetic oil. Fluid drives can be grouped into four categories: (1) hydrokinetic, (2) hydrodynamic, (3) hydroviscous, and (4) hydrostatic. The hydrokinetic, hydrodynamic, and hydroviscous drives are all slip-type devices.

The hydrokinetic fluid drive, commonly referred to as a fluid coupling, consists of a vaned impeller connected to the driver and a vaned runner connected to the load. The oil is accelerated in the impeller and then decelerated as it strikes the blades of the runner. Thus there is no mechanical connection between the input and output shafts. Varying the amount of oil in the working circuit changes the speed. This provides infinite variable speed over the operating range of the drive. Figure 6.19 is a representation of such a drive.

The circulating pump, driven from the input shaft, pumps oil from the reservoir into the housing through an external heat exchanger and then back to the working elements. The working oil, while it is in the rotating elements, is thrown outward where it takes the form of a

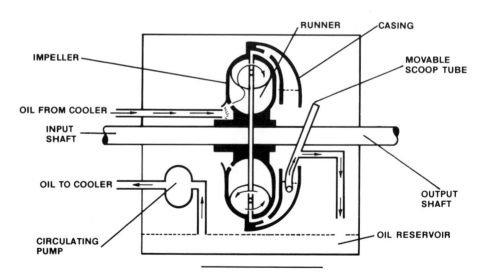

Fig. 6.19 Diagram of a hydrokinetic fluid drive. (Courtesy of Industrial Products Division, American-Standard, Dearborn.)

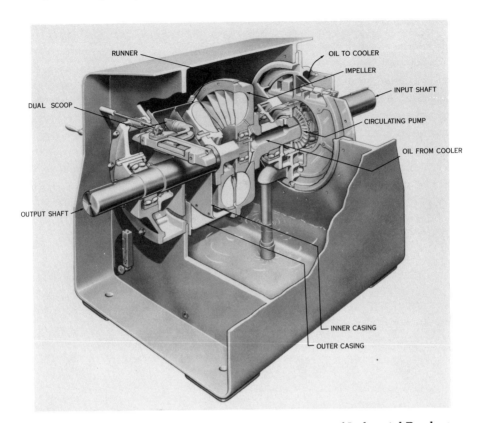

Fig. 6.20 American-Standard fluid drive. (Courtesy of Industrial Products Division, American-Standard, Dearborn.)

toroid in the impeller and runner. Varying the quantity of oil in this toroid varies the output speed. A movable scoop tube controls the amount of oil in the toroid. The position of the scoop tube can be controlled either manually or with automatic control devices. The scoop tube adjustment gives a fast response and smooth stepless speed control over a wide speed range, i.e., 4 to 1 with a constant-torque load and 5 to 1 with a variable-torque load. In addition to providing speed control, the fluid drive limits torque and permits no load starting on high-inertia loads.

These units range in size from 2 to 40,000 hp, as illustrated in Fig. 6.20.

Efficiency

The fluid drives have two types of losses:

Circulation Losses. These losses include friction and windage losses, the power to accelerate the oil within the rotor, and the power to drive any oil pumps that are part of the system. These losses are relatively constant and are approximately 1.5 percent of the unit rating.

Slip Losses. As in the case of eddy current couplings, the torque at the input shaft is equal to the torque required at the output shaft:

$$\text{Load (output) hp} = \frac{\text{rpm}_2 T_L}{5250}$$

$$\text{Motor (input) hp} = \frac{\text{rpm}_1 T_L}{5250}$$

$$\begin{aligned} \text{Slip loss} &= \text{motor hp} - \text{load hp} \\ &= \frac{T_L(\text{rpm}_1 - \text{rpm}_2)}{5250} \end{aligned}$$

where
rpm_1 = coupling input speed (motor)
rpm_2 = coupling output speed (load)
T_L = load torque, ft-lb

The slip efficiency is then

$$\frac{\text{Input hp} - \text{slip loss}}{\text{input hp}} \times 100$$

$$= \frac{(T_L \text{rpm}_1/5250) - (T_L/5250)(\text{rpm}_1 - \text{rpm}_2)}{T_L \text{rpm}_1/5250}$$

$$= \frac{\text{rpm}_2}{\text{rpm}_1}$$

$$\text{Total input hp} = \frac{\text{output hp}}{\text{slip eff}} + \text{circulation hp losses}$$

$$\text{Coupling efficiency} = \frac{\text{output hp}}{\text{input hp}}$$

Figure 6.21 illustrates the typical performance of a fluid coupling driving a load whose horsepower requirement varies as the speed cubed. The maximum speed of the fluid drive at full load is about 98

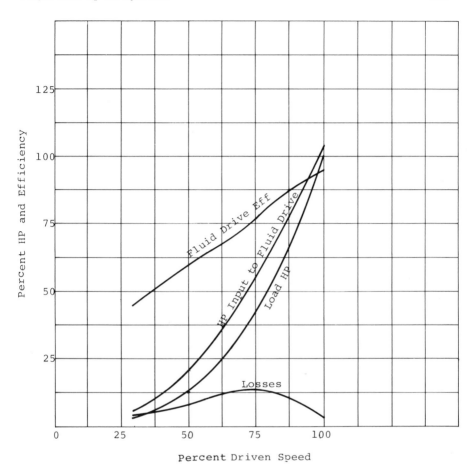

Fig. 6.21 Fluid coupling variable-speed drive characteristics when driving a load that varies as the speed cubed.

percent of the driving motor speed, and with circulation losses of 1.5 percent the maximum efficiency is 96.5 percent at maximum speed.

Figure 6.22 is a typical installation of a fluid drive system driving a centrifugal pump.

Hydrostatic Drives

A hydrostatic variable-speed drive consists of a positive displacement hydraulic pump driven by an induction motor, a positive displacement hydraulic motor, and necessary hydraulic controls. The

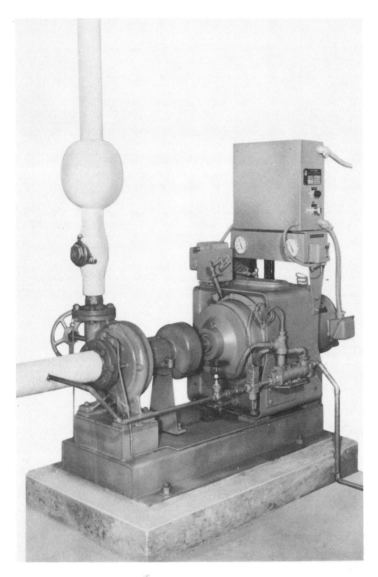

Fig. 6.22 Installation of an American-Standard fluid drive driving a centrifugal pump. (Courtesy of Industrial Products Division, American-Standard, Dearborn.)

hydraulic pump and motor are usually separate units. However, this type of drive is also offered as a package consisting of the hydraulic pump, piping, and the hydraulic motor mounted in a common housing.

When the hydraulic pump is driven by a constant-speed AC induction motor, the variable output is obtained by controlling the speed of the hydraulic motor. Commonly, the easiest system to design may be the most energy inefficient. Throttling any valve in the hydraulic system generates heat and consumes energy. The significance of this power loss is expressed as follows:

$$\text{Power loss (hp)} = \frac{\text{pressure drop (psi)} \times \text{flow (gpm)}}{1714 \times \text{pump efficiency}}$$

The most efficient hydraulic system is one that has no valves. However, such a system will also have very limited speed control. Many methods of control have been developed for hydraulic systems, and the method used depends on the types of pump and motor used and the characteristic of the load. Many of the systems are used on mobile equipment and machine tools, but they are not generally cost effective on industrial applications such as pumps and fans.

AC Variable-Frequency Drives

The squirrel-cage induction motor is normally considered a constant-speed device with an operating speed 2 to 3 percent below its synchronous speed. However, efficient operation can be obtained at other speeds if the frequency of the power supply can be changed. The synchronous speed of an induction motor can be expressed by

$$Ns = \frac{120f}{p}$$

where
   Ns = synchronous speed, rpm
    f = power supply frequency, Hz
    p = number of poles in motor stator winding

A four-pole induction motor which when operated on a 60-Hz power supply has a synchronous speed of 1800 rpm will operate at the following synchronous speeds as the power supply frequency is changed:

Power frequency (Hz)	Motor synchronous speed (rpm)
120	3600
90	2700
60	1800
30	900
15	450
7.5	225

Variable-Frequency Power Supplies

Recent advances in power semiconductor technology have provided power conversion systems that can supply variable frequency power to accomplish the preceding frequency changes and hence variable-speed operation of three-phase induction motors. These power supplies are commonly referred to as inverters. This nomenclature is somewhat misleading since the units actually consist of an AC to DC converter section which converts the incoming 60-Hz AC power to DC power and a DC to AC inverter section which inverts the DC power to the AC adjustable-frequency output power.

For adjustable-speed operation of an AC induction motor supplied by an inverter, the output voltage is varied proportional to the frequency except at low frequencies, where the voltage is boosted above its proportional level. The torque developed by an induction motor is proportional to the magnetic flux in the motor air gap. This magnetic flux is directly proportional to the applied voltage and inversely proportional to the applied frequency. Therefore, as the output frequency (and motor speed) is reduced, the applied voltage must also be reduced to prevent magnetic saturation and excessive motor losses.

This ratio of voltage to frequency applies to the net voltage required to develop the magnetic flux in the motor. At normal frequencies, the stator winding resistance drop is only a small percentage of the applied voltage. However, since this resistance drop remains constant as the frequency is reduced, at very low frequencies the resistance drop becomes a large percentage of the applied voltage. Therefore, it is necessary for the inverter to provide voltage boosting at the low frequencies to compensate for this resistance drop, as illustrated in Fig. 6.23, and to maintain the motor torque at low frequencies.

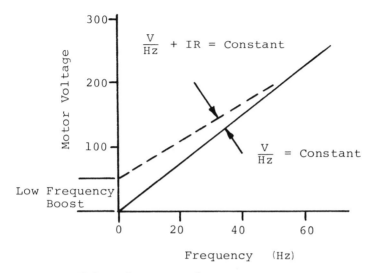

Fig. 6.23   Voltage-frequency relation.

There are a number of types of inverters available, and the two most common types to provide this frequency conversion are the six-step method and the pulse width modulated method.

Figure 6.24 illustrates the basic functions and output of the six-step inverter. In this system, the 60-Hz input voltage is rectified to adjustable voltage DC by means of an adjustable voltage rectifier bridge. Then each of the three-phase output lines is switched from positive to negative for 180° of each 360° cycle. The phases are sequentially switched at 120° intervals, thus creating the "six-step" line to neutral voltage shown in Fig. 6.24. Output voltage is controlled by controlling the voltage level of the DC bus. This is normally achieved either by a controlled rectifier input or by a switching-type chopper regulator. The output frequency is controlled by a reference signal which sets the control logic to achieve the correct gate or base-drive signals for the SCRs or transistors in the inverter section.

The six-step system produces fairly high fifth, seventh, eleventh, and thirteenth harmonic currents. These harmonic currents do not contribute to the output torque but do result in increased losses in the motor, thus decreasing the efficiency and increasing the motor heating.

The current pulsations from the six-step inverter produce torque pulsations in the motor. At high speeds, above about 200 rpm, the rotor inertia smooths out these pulsations. However, at low speeds, for example, on a four-pole induction motor operating at 7 Hz and 200

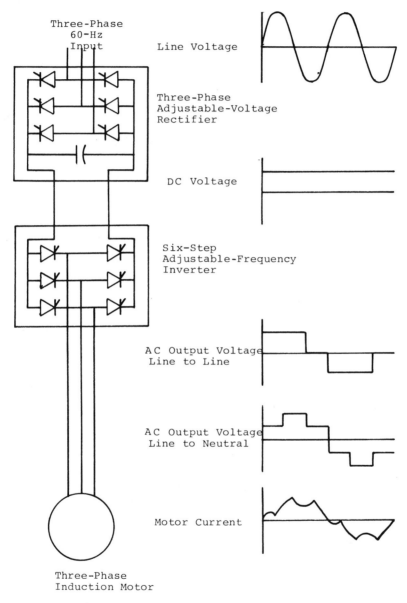

Fig. 6.24   Basic six-step inverter system.

rpm, the torque pulsations can cause a pulsation in speed or *cogging*. This can be a problem for applications requiring smooth output at very low speeds.

The pulse width modulated (PWM) system rectifies the 60-Hz line voltage to produce a constant potential DC voltage. The DC voltage is then applied to the motor in a series of pulses, the widths of which are varied to control frequency, voltage, and harmonic content of the output waveform, as illustrated in Fig. 6.25.

By selecting the width and spacing of the pulses, lower-order harmonics, such as the fifth, seventh, and eleventh, can be eliminated in the waveform. If the pulse rate is high enough, the motor inductance presents a high impedance so that the pulse-rate-frequency current is insignificant. From the motor viewpoint, it is desirable to have a high-frequency pulse rate. From the inverter viewpoint, since most of the losses occur during switching, it is best to have a low pulse rate. However, the number of pulses per cycle must be maintained high enough to avoid troublesome harmonics which may be resonant with the motor components. The switching and recovery time on SCRs tends to limit their use on PWM systems. Power transistors or gate-turnoff (GTO) devices have much faster switching times, so they can be used at high pulse rates, but the use of these devices is limited by the current and voltage ratings presently available. While the PWM inverters improve the waveforms by eliminating the low-order harmonics, they impose a series of high-voltage impulses on the motor winding. Although the winding inductance smooths the current waveform, the rapid voltage changes produce insulation stresses on the first few turns of each of the motor windings. Full-voltage PWM systems produce the most severe stresses, particularly at low speeds where the motor back-EMF is low.

There are numerous variations of these two types of adjustable-frequency inverters, but the principle of operation is essentially the same. As with any relatively new product, changes and improvements are being accomplished every day. The improvements come primarily from the increasing use of integrated circuits as well as microprocessors, which have greatly reduced the number of control logic components. Also the use of power transistors and GTO SCRs has reduced the cost of power elements on lower-horsepower units. These, plus improved designs and techniques, have reduced and will continue to reduce the size and cost of the inverters. At the same time, the performance and reliability have been improved.

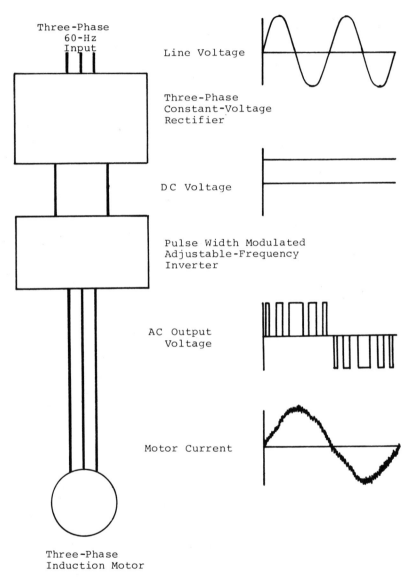

Fig. 6.25 Basic pulse width modulated inverter system.

Motor Characteristics

When the inverter output has a constant voltage to frequency ratio plus a low-frequency voltage boost to compensate for winding resistance, the motor will be capable of producing full-rated torque at the rated current over the entire speed range and with a constant slip speed. That is, if a four-pole motor operates at 1750 rpm with full-load torque at 60 Hz, then it will operate at 850 rpm with the same loads and same current at 30 Hz and 120 rpm at 6 Hz.

Basically, the motor speed-torque curve, as shown in Fig. 6.26,

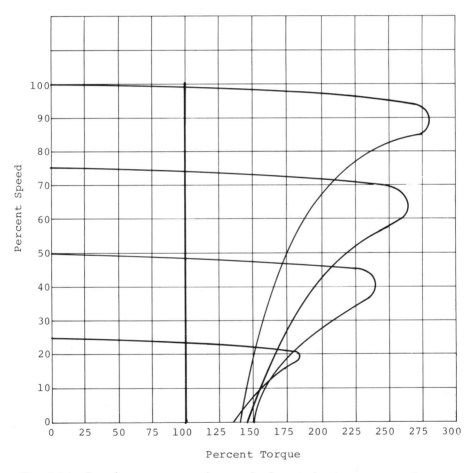

Fig. 6.26 Speed-torque curves for a polyphase induction motor with constant volts per cycle and a sinusoidal input.

retains its shape but moves across the speed axis as frequency is changed. This characteristic enables a motor to be brought up to speed while developing full-load torque with only the rated current by slowly raising the frequency and voltage.

In many inverters, the voltage to frequency ratio is maintained from the minimum frequency up to the frequency where maximum voltage from the DC bus is required. This point is usually the base speed. Above this point, frequency is increased with the voltage held constant. Since motor flux is proportional to the voltage to frequency ratio, further increases in frequency reduce the flux, and hence the available torque decreases as the square of frequency increases.

By operating the motor at a higher slip, it is possible to develop constant-horsepower output up to about 150 percent of the base speed, as shown in Fig. 6.27. At higher speeds, the available power drops off rapidly.

Constant Torque

If the driven load has a constant-torque characteristic for a change in speed, then as speed is decreased by reducing frequency and voltage, the motor current remains essentially constant. Motor losses also remain constant, but the effectiveness of the motor cooling fan decreases. As a result, the motor winding temperature increases; if it goes high enough, it can cause irreversible damage to the insulation.

The basic problem is that most inverters are designed to operate with standard 60-Hz motors, and the optimum base frequency may be some other value, depending on load requirements. Where the economics or the application justifies a custom motor, the motor can be wound for the voltage and frequency that provide optimum performance for the complete system.

Variable Torque

In many applications, the driven load does not require constant torque. A fan, for example, has a square law torque curve; torque decreases with the square of the speed. Here, required motor torque is very low at low speeds. To keep the current low and avoid overheating, the voltage to frequency ratio must be reduced at low frequencies.

This characteristic can be designed into the inverter, and some off-the-shelf units provide it. Even though the load is light at low speeds, the magnetizing current may still be high enough to overheat

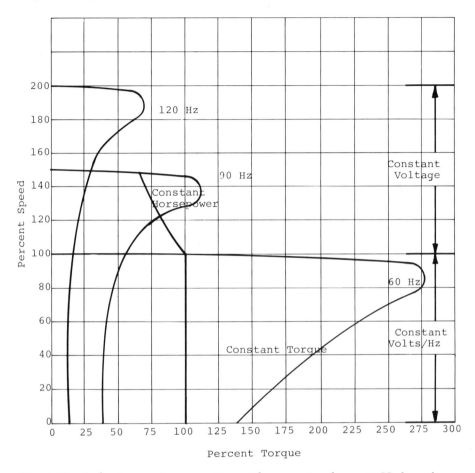

Fig. 6.27 Induction motor operation at frequencies above 60-Hz base frequency.

the motor. For such loads, the frequency boost should be set at minimum.

High Starting Torque

If the load requires a high starting torque, we need to use a drive with a relatively high low-frequency voltage boost. Such a high boost may cause overheating if the motor is run for long periods at low speeds, but this can be overcome by using an oversized motor.

Alternatively, the inverter can be programmed to provide high boost for starting and reduced boost for running, but this type of control is not found on standard units. Ideally the inverter will have a custom-designed microprocessor control that adjusts the voltage to frequency ratio for maximum efficiency at all speeds and torques.

System Efficiency

The efficiency of inverters is usually quoted at 90 to 95 percent at the base frequency, which is typically 60 Hz. At either lower or higher frequencies, efficiency is normally lower.

Motor efficiency is lower with an inverter supply than with a sine-wave supply because of harmonic current losses. Motor efficiency is further reduced as speed is reduced because of the increasing effect of resistance losses. Thus an inverter with a nominal efficiency of 95 percent supplying a motor with a rated sine-wave efficiency of 90 percent gives an overall efficiency between 70 and 80 percent, depending on the specific waveform supplied. Even this overall efficiency drops as frequency is changed from the nominal 60 Hz. Figures 6.28 and 6.29 illustrate the typical components and system efficiencies for an inverter system.

Most inverter manufacturers state that any standard induction motor can be used with their units. However, all inverters produce motor losses exceeding those found in normal 60-Hz operation, reducing motor efficiency and motor life to some extent. Continuous operation at low frequencies can be especially severe. Figure 6.30 shows the increase in motor winding temperature on a constant-torque application as the motor speed is decreased.

A four-pole induction motor running at design speed (typically 1750 rpm) and powered by a six-step inverter normally has a temperature about 20 percent hotter than it would if it were powered by a sine-wave supply. Most currently available motors—particularly the high-efficiency models—can tolerate this increased heating without reducing insulation life significantly.

If energy savings are the main justification for the drive, this overall system efficiency is the controlling parameter, not the individual rated efficiencies of the motor and inverter.

AC Variable-Frequency Drive Application Guide

Unfortunately, the selection and application of an AC variable-frequency induction motor drive system are more complex than the selection of a fixed-speed induction motor. The duty cycle of the in-

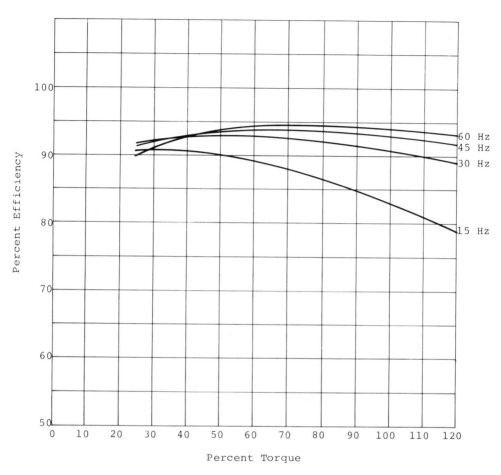

Fig. 6.28 Typical efficiency curves for a polyphase induction motor with constant volts per cycle and a sinusoidal motor input.

verter-motor combination must be checked at all load conditions to make certain that the particular drive combination is suitable for the given application. In addition, some applications may require control options such as digital speed control, closed-loop speed control, frequency metering, variable voltage boost, or remote signal inputs such as pressure or temperature. The characteristics that determine the appropriate drive combination are the following:

1. Speed range required
2. Speed-torque characteristics of the load
3. Load inertia

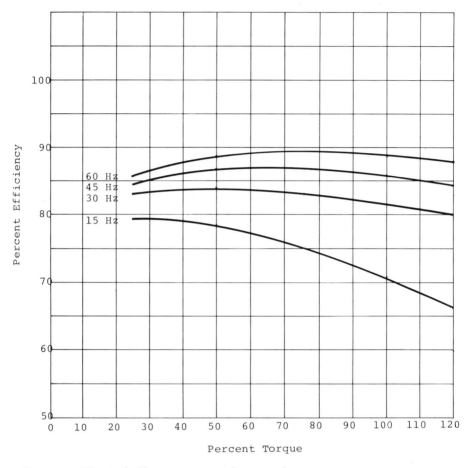

Fig. 6.29 Typical efficiency curves for an induction motor-inverter system with constant volts per cycle and a nonsinusoidal motor input.

4. Load acceleration and deceleration times
5. Required operating time at various speeds
6. Inverter output waveform and its approximate harmonic content
7. System efficiency over the operating range
8. Regenerative energy dissipation in the inverter
9. Motor temperature rise at the required duty cycle and the voltage to frequency ratio provided by the inverter
10. Motor rating based on the duty cycle
11. Motor insulation-life derating for its input waveform (applies to full-voltage PWM systems)

12. Inverter construction and enclosure
13. Motor enclosure

Wound Rotor Motor Drives with Slip Loss Recovery

The wound rotor motor has normally been used for short-time duty applications such as cranes and hoists where torque control was of prime importance. When it has been used on continuous duty installation, the major purpose was to obtain controlled starting and acceleration. The reason for this limited use was that high slip losses occur at

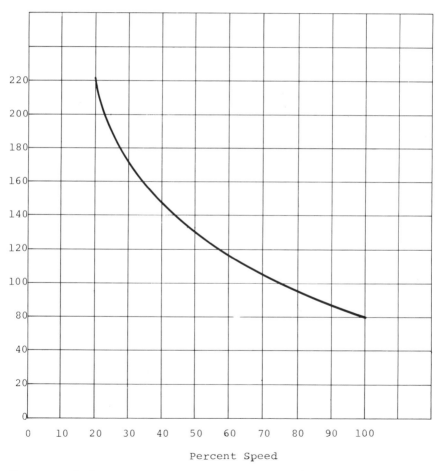

Fig. 6.30 Induction motor temperature rise (°C) for an inverter power source with a constant-torque load and constant volts per cycle.

speeds below normal operating speed. With the development of power electronics and solid-state inverters, systems have been developed to recover these slip losses.

A typical wound rotor solid-state adjustable-speed control system to accomplish the recovery of slip losses is the Econopak™ manufactured by the Square D Company. Figure 6.31 is the circuit diagram for this system. The secondary side of the motor is connected to a three-phase rectifier bridge. The output of the bridge is connected to a fixed-frequency inverter the output of which is connected to the power supply. The effective rotor resistance and hence the motor speed are controlled by controlling the firing angle of the power SCRs in the inverter. At the same time, the rotor slip losses are returned to the power supply instead of being dissipated in external resistors. In this particular control system, when operating at full speed, the shorting thyristors short the wound rotor secondary. No current flows through the inverter at this time, and the efficiency is that of the motor running across the line. Figures 6.32 and 6.33 show a 200-hp, three-phase, 60-Hz Econopak™ unit.

The speed range that can be obtained is determined by the motor secondary (rotor) voltage; for instance, for a 100 percent speed range system,

480-V power supply: The rotor voltage must be 380 V or less.
600-V power supply: The rotor voltage must be 480 V or less.

and for a 50 percent speed range system,

480-V power supply: The rotor voltage must be between 600 and 760 V.
600-V power supply: The rotor voltage must be between 750 and 960 V.

For power supply voltages above 600 V, such as 2300 and 4160 V, the motor primary can utilize the line voltage. However, a matching transformer is required at the output of the rotor inverter, and it need only be large enough to handle the rotor losses, not the total motor input. Power factor correction capacitors are supplied with the Econopak™ controller. The capacitors are sized to correct the line lagging power factor at full speed to 75 percent. At lower speeds with a cubic-type load such as a pump or fan, the load power factor will improve and become leading at 72 percent speed. At 60 percent speed, the power factor will be 70 percent leading, and the capacitors will be switched out of the circuit.

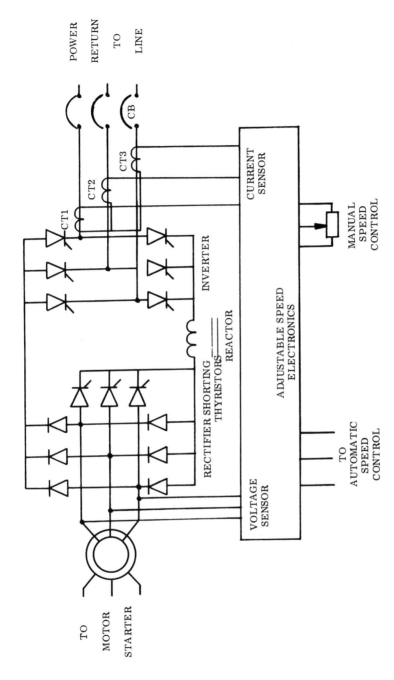

Fig. 6.31 Econopak™ power circuit diagram. (Courtesy of Square D Company, Palatine, Ill.)

Fig. 6.32 Square D 200-hp Econopak™ control. (Courtesy of Square D Company, Palatine, Ill.)

The efficiency of the controller is approximately 98.5 percent and is constant over the speed range; thus the system is very efficient in recovering the slip losses and raising the system efficiency.

Consider a 200-hp wound rotor motor on a pumping installation where the motor horsepower load is a cubic function of the speed.

Without the slip recovery controller at full speed (1764 rpm):
Horsepower output, 200 hp
Motor efficiency, 94%

At one-half speed (882 rpm):
   Horsepower output, 25 hp
   Motor efficiency, 46%
With the slip recovery controller at full speed (1764 rpm):
   Horsepower output, 200 hp
   Motor efficiency, 94%
   Overall system efficiency, 94%
At one-half speed (882 rpm):
   Horsepower output, 25 hp

Fig. 6.33 Square D 200-hp Econopak™ control. (Courtesy of Square D Company, Palatine, Ill.)

Motor efficiency, 46%
Overall system efficiency, 84%

Note that recovery of the slip losses at one-half speed increased the efficiency from 46 to 84 percent.

The wound rotor motor with a slip recovery system used on a pump or fan application can usually be operated over a 50 percent speed range with self-ventilation. For applications requiring continuous operation below 50 percent speed, forced ventilation may be required for the motor.

This type of drive system offers an energy-efficient system comparable to adjustable-frequency systems and superior to slip-loss-type systems.

6.3   APPLICATION OF ADJUSTABLE-SPEED SYSTEMS TO FANS

Various types of fans used to move air or other types of gases are one of the largest consumers of electric power and one of the largest users of integral horsepower electric motors. In general, fans can be divided into two broad categories: centrifugal fans and axial flow fans.

The guide for selection of the proper fan for a given application is covered by most fan manufacturer catalog information. However, to obtain the most energy-efficient system, it is necessary to examine the methods to control airflow in a given system.

First, it is necessary to determine the system resistance characteristic for various airflow rates. A curve can be developed for the system in terms of the air volume versus the static pressure required. This curve generally follows a simple parabolic law in which the static pressure or resistance to airflow varies as the square of the volume of air required. Figure 6.34 shows the system curve for a specific example that will be discussed later.

Next is the selection of the type of fan. Many of the applications in general heating, ventilating, and air conditioning systems involve centrifugal fans. These fans generally fall into three major categories based on the type of impeller design:

Backward-Curved Blades.   The horsepower reaches a maximum near peak efficiency and becomes lower toward free delivery. Figure 6.35 is the typical performance curve for the backward-curved

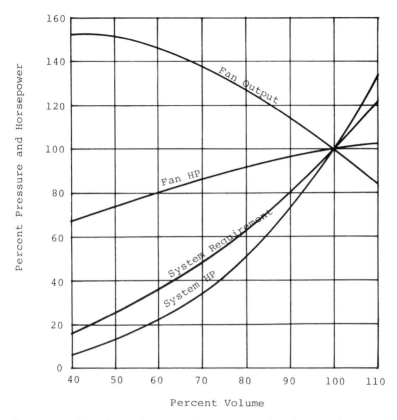

Fig. 6.34 Sample application, characteristic fan data, and system characteristic.

fan. The volume is the percent of free flow volume, and the pressure is the percent of static pressure at zero volume. The horsepower is the percent of maximum horsepower.

Radial Blades. These have higher pressure characteristics than the backward-curved fan. The horsepower rises continually to free delivery. Figure 6.36 is the typical performance curve for radial blade fans.

Forward-Curved Blades. The pressure curve is less steep than that for the backward-curved fan. The peak efficiency is to the right of the peak pressure. The horsepower rises continually to free delivery. Figure 6.37 is the typical performance curve for forward-curved fans.

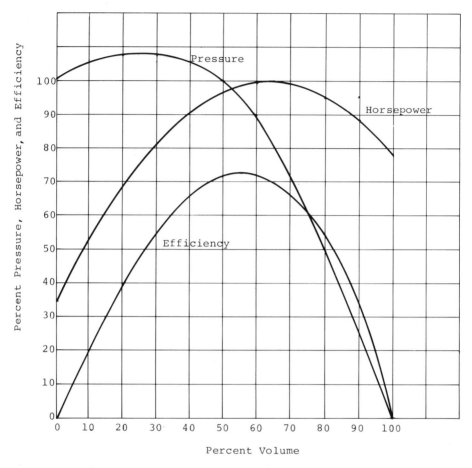

Fig. 6.35 Characteristic curves for the backward-curved fan.

The size and type of fan selected should be such that the fan is operating near its peak static efficiency for the maximum flow rate required. The performance of the fan at other speeds will follow the following fan laws:

1. The volume V of air varies as the fan speed.
2. The static pressure P varies as the square of the fan speed.
3. The horsepower varies as the cube of the fan speed.

$$\text{Static efficiency} = \frac{VP}{6369 \times hp}$$

Several methods of controlling the airflow can be considered:

Damper control
Variable inlet vane control
Hydrokinetic or fluid drive systems
Eddy current drive systems
Mechanical variable-speed drive units
AC variable-frequency systems
Wound rotor motor with slip recovery systems
Two-winding, two-speed motors

The comparison of these systems is best illustrated by an example.

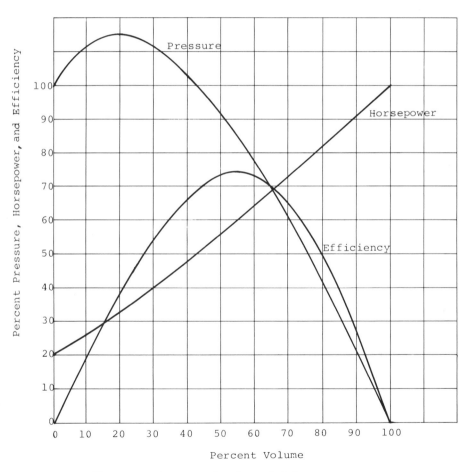

Fig. 6.36 Characteristic curves for the radial fan.

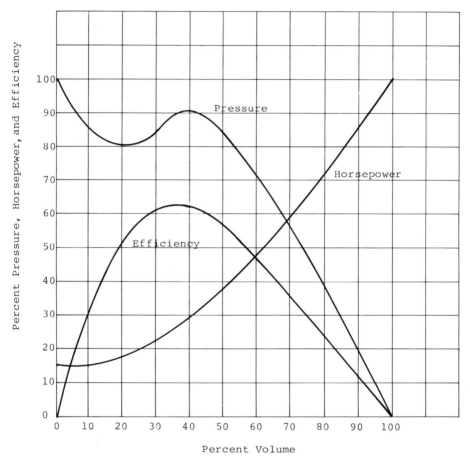

Fig. 6.37   Characteristic curves for the forward-curved fan.

EXAMPLE

The characteristic curves of the fan and air system for the example are shown in Fig. 6.34. The data have been shown in percentages of the pressure, volume, and horsepower at the balance point where the system resistance curve and the fan curve intersect.

Based on this fan-system curve, the horsepower input at the fan to produce the airflow is also shown. Figure 6.38 shows the input horsepower required for each of the preceding systems compared to the fan horsepower required as the airflow requirement is varied between 40 and 100 percent volume.

To determine the most energy-efficient system, the operating cycle, i.e., the percent time operating at each volume flow, must be determined; then the energy savings and net present worth of each system can be compared to the most inefficient system, i.e., one using a discharge damper.

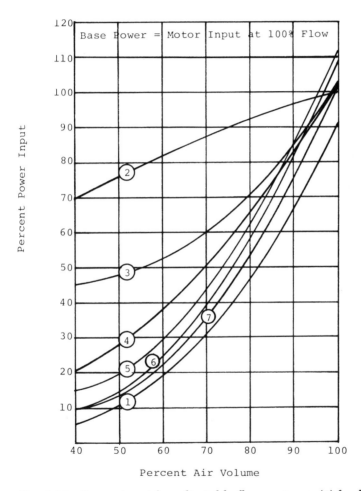

Fig. 6.38 Power input for adjustable-flow systems: (1) fan hp required, (2) damper control, (3) inlet vane control, (4) hydrokinetic and eddy current system, (5) mechanical varispeed system, (6) variable-frequency system, (7) wound rotor with rotor loss recovery system; 100 percent power is the power input to the motor at 100 percent airflow.

To illustrate these data, three different operating cycles have been assumed with the following data:

Horsepower at full air volume, 100 hp
Annual operating hours, 4000 hr
Initial power rate, $0.06/kWh
Annual increase in power rate, 10 percent
Cost of money, 15 percent
Tax rate, 40 percent
System life, 10 yr

The operating cycles are as follows:

Time at each air volume (%)

Cycle	100% air vol.	80% air vol.	60% air vol.
I	75	15	10
II	33	33	34
III	10	15	75

Two other operating cycles based on fixed-speed operation at two speeds using a two-winding, two-speed motor with an 1800/1200-rpm speed combination are as follows:

operating cycle 1-A:
   100 percent (full) air volume: 75 percent of the time
   66 percent air volume: 25 percent of the time
operating cycle 1-B:
   100 percent (full) air volume: 25 percent of the time
66 percent air volume: 75 percent of the time

The summary of the annual power savings in kilowatt-hours for each system is shown in Table 6.2. The net present worth of the energy savings based on the defined assumptions is shown in Table 6.3. This summary shows the influence of the operating cycle on the power savings and the net present worth of those savings. When compared to the first cost of each system in conjunction with other considerations, including reliability, flexibility, maintenance, and environment, this net

Table 6.2 Comparison of Annual kWh Savings for Adjustable Flow Fan Systems[a]

Operating cycle	System					
	Multispeed motor	Variable vane inlet	Fluid drive or eddy current drive	Mechanical varidrive	Variable-frequency system	Wound rotor with slip recovery
I	—	10,431	15,080	4,070	12,550	32,768
II	—	51,515	71,916	81,355	87,879	103,008
III	—	80,382	118,948	141,854	148,216	157,834
I-A	39,396	—	—	—	—	—
III-A	129,060	—	—	—	—	—

[a]Base for comparison: damper control.

Table 6.3 Net Present Worth Comparison in Dollars for Adjustable-Flow Fan Systems[a]

Operating cycle	System					
	Multispeed motor	Variable vane inlet	Fluid drive or eddy current drive	Mechanical varidrive	Variable-frequency system	Wound rotor motor with rotor loss recovery
I	—	3,558	5,143	1,388	4,271	10,176
II	—	17,571	24,528	27,746	29,973	35,029
III	—	27,416	40,569	48,381	50,551	53,831
I-A	13,727	—	—	—	—	—
III-A	44,018	—	—	—	—	—

[a]Base for comparison: damper control.

present worth will provide an economic basis for selecting the most cost-effective system.

6.4 APPLICATION OF ADJUSTABLE-SPEED SYSTEMS TO PUMPS

Pumps are the largest user of electric motors in the integral horsepower sizes 1 hp and larger. The selection and application of electric motors and adjustable-speed systems to pumps become very complex and difficult because of the large numbers of types of pumps and their lack of standardization. The pumps fall into two broad categories: displacement pumps and dynamic pumps. The dynamic pumps include noncentrifugal types and centrifugal types. The highest percentage of the pumps used for industrial processes are of the centrifugal type; therefore the discussion in this section will be limited to drive applications of this type.

In the selection of a pump for a given system, the system characteristics must be determined in terms of flow rate in gallons per minute versus total head in feet under all flow rates expected. The system should include an allowance for pipe corrosion and other factors that affect the system characteristics. The pump then selected should be sized to the system characteristic at the maximum flow rate such that the efficiency is close to the optimum for the pump. Then for an adjustable-speed system, it is necessary to check the pump performance at the minimum flow point required.

The various pump manufacturers provide data for their pumps and also provide assistance in selecting the correct pump for the application. The initial guide is a performance curve at a fixed speed, as shown in Fig. 6.39, indicating the best efficiency point on each pump in a family of pumps. For a given pump, detail performance curves such as those in Fig. 6.40 are available from the manufacturer.

The following relationship applies to pumps (as discussed for fans in Sec. 6.3):

$$\frac{Q_1}{Q_2} = \frac{N_1}{N_2}$$

$$\frac{H_1}{H_2} = \left(\frac{N_1}{N_2}\right)^2$$

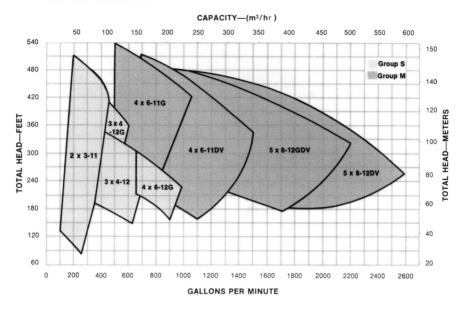

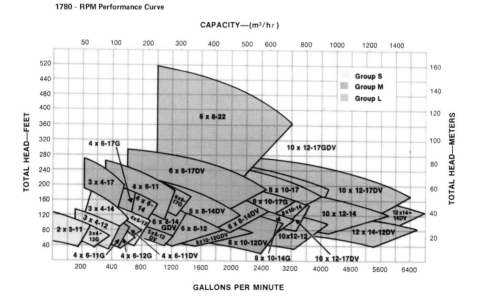

Fig. 6.39 3550- and 1780-rpm centrifugal pump performance curves. (Courtesy of Gould's Pumps Inc., Seneca Falls, N.Y.)

168

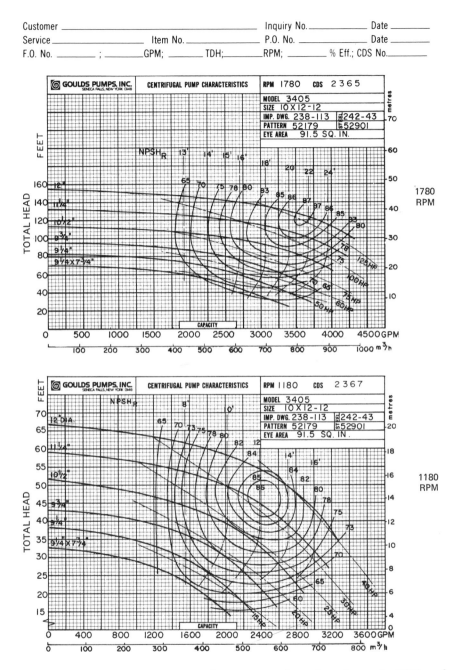

Fig. 6.40 Centrifugal pump performance at the selected speeds 1780 and 1180 rpm. (Courtesy of Gould's Pumps Inc., Seneca Falls, N.Y.)

$$\frac{BHP_1}{BHP_2} = \left(\frac{N_1}{N_2}\right)^{3.}$$

$$BHP = \frac{QH \times Sp.\ Gr.}{3960 \times Pump\ Eff}$$

$$Pump\ Eff = \frac{QH \times Sp.\ Gr.}{3960 \times BHP}$$

where

Q = capacity, gpm
H = total head, ft
BHP = brake hp
N = pump speed, rpm
Sp. Gr. = specific gravity of liquid

To illustrate the application and cost analysis of various methods of flow control, consider an example comparing the following:

Throttling in the discharge line
Eddy current or hydraulic fluid drive systems
AC variable-frequency system
Wound rotor motor with slip recovery system

Figure 6.41 shows the system flow characteristic and the centrifugal pump characteristics at various pump speeds for the specific application. Figure 6.42 shows the pump characteristic at various speeds and the system characteristics as a function of flow requirements.

The pump BHP, the power input, and the power losses from the power line to the pump can be calculated for each method of flow control. A summary of these calculations is shown in Table 6.4 for three flow conditions: full flow (100 percent), 75 percent flow, and 50 percent flow.

The annual power savings and present net worth of the savings for each system can be determined for a specific set of conditions. To continue the sample calculations, set the following conditions:

Product life or life cycle, 7 yr
Annual operating hours, 8000 hr
Operating cycle: full flow, 50 percent of the time
    75 percent flow, 30 percent of the time
    50 percent flow, 20 percent of the time

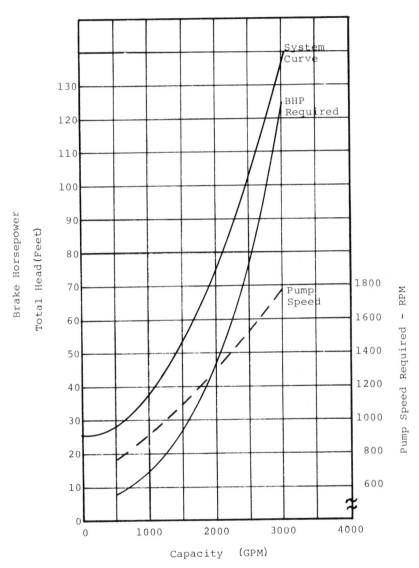

Fig. 6.41 Sample fluid system characteristic.

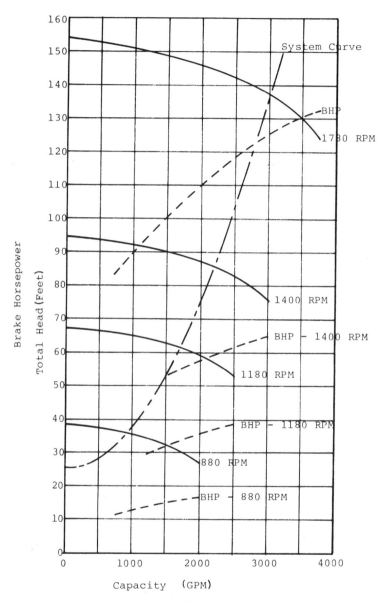

Fig. 6.42 Centrifugal pump characteristics.

Table 6.4 Summary of Sample Calculations for Adjustable-Flow Systems

Flow rate (%)	Flow (gpm)	Pump[a] BHP required	Throttling system		Eddy current drive or fluid drive		AC variable frequency system		Wound rotor motor with slip loss recovery	
			Input (kW)	Losses (kW)	Input (kW)	Losses (kW)	Input (kW)	Losses (kW)	Input (kW)	Losses (kW)
100	3000	125	100.81	7.56	107.26	14.0	103.96	10.71	101.91	8.66
75	2250	60	92.65	47.89	61.29	16.5	51.15	6.39	51.45	6.69
50	1500	28	80.56	59.67	36.89	16.0	24.60	3.72	25.17	4.28

[a]For adjustable-speed pump.

Table 6.5 Annual Energy Savings and Net Present Worth for Adjustable-Speed Pumping Systems[a]

System	Annual savings (kWh)	Net present worth ($)
Fluid drive or eddy current system	120,056	30,907
AC variable-frequency system	179,536	45,448
Wound rotor with slip recovery system	183,104	47,141

[a]Base for comparison: throttling system.

Full-flow BHP, 125 hp
Initial power rate, $0.06/kWh
Annual increase in power rates, 15 percent
Cost of money, 20 percent
Tax rate, 40 percent

The summary of these calculations in Table 6.5 shows the savings that can be achieved by means of adjustable-speed pumping systems compared to a throttling-type system. These calculations indicate that substantial power savings can be achieved with variable-speed-type pumping systems. The selection of the most effective system depends on the comparison of the system cost to the net present worth of the system savings and application factors such as environment and maintenance which influence the system selection.

7

Economics of Energy-Efficient Motors and Systems

7.1 GENERAL REVIEW

The method of determining the economics of the improvement in an electric motor or system efficiency can be either a simple payback calculation or a more comprehensive calculation depending on the user's requirements. In any method used, the essential factors that are required include the following:

Efficiencies of motors or systems being compared
Hours of operation per year
Power costs in dollars per kilowatt-hour
Motor or system loading

It is essential that the efficiencies used be comparable. That is, the nominal efficiencies are used for the motors being compared or the minimum efficiencies are used for the comparison, and the efficiencies are determined by equivalent test methods. Earlier discussion has pointed out the discrepancies that can occur because of efficiency values and test variations. Similarly, the method of payback analysis used should be consistent in order to obtain comparable economic data.

In the case of comparing systems such as a system consisting of a motor and eddy current drive connected to a load or a variable-frequency system, the system efficiency is the product of the component efficiencies. This system efficiency is the efficiency that must be considered in determining the total energy consumption by each system in order to make an adequate comparison of the energy costs.

7.2 LIFE CYCLE

In those analysis methods based on the life cycle, the question is, What is the life cycle? This may be determined by various criteria:

Operating life of the electric motor
Operating life of the driven equipment
Operating life of the process

If the operating life or projected life of the driven equipment or the process is 5 to 10 yr or less, this life can be used for any life-cycle calculations. If the projected life is over 10 yr, then a basis for the motor life needs to be determined. There is no established design life for electric motors since there are many factors that influence the motor life. However, Table 7.1 indicates the average motor life for three-phase motors. This table was based on a study made at the U.S. Department of Energy. Unless other data are available, this table can be used as a base for life-cycle calculation with appropriate adjustments for adverse operating conditions.

The major factor in the electric motor life is the life of the insulation system.

The application factors that influence the insulation life and hence the motor life are the following:

Table 7.1 Average Electric Motor Life   3 Phase motors

hp range	Average life (yr)	Life range (yr)
Less than 1	12.9	10–15
1–5	17.1	13–19
5.1–20	19.4	16–20
21–50	21.8	18–26
51–125	28.5	24–33
Greater than 125	29.3	25–38
Average all units = 13.27 yr		

Source: DOE Report DOE/CS-0147, 1980.

Loading relative to rated output

Operating hours per year

Ambient conditions: temperature, relative humidity, and dirt and
    contamination

Voltage conditions

The influence of these factors on motor life and motor failure is shown
in Table 7.2. It is generally recognized that most insulation systems age
in accordance with a chemical reaction rate theory based on the Ar-
rhenius model. Thus, the insulation life can be expressed as

$$\log L = \frac{B}{T} - A$$

where

L = insulation life, hr of operation

Table 7.2 Motor Failure Survey by a Large
Service Shop[a]

Cause of failure	Total failure (%)
Overload (overheating)	25
Normal insulation deterioration (old age)	5
Single phasing	10
Bearing failures	12
Contamination	
Moisture	17
Oil and grease	20
Chemical	1
Chips and dust	5
Total	43
Miscellaneous	3

[a]Based on the study of 4000 failures over several years.

Source: J. C. Andreas, Automation, April 1974, pp.
82 and 85.

T = absolute temperature, = °C + 273

B, A = constants for a particular insulation system

The value of the constants B and A is a function of the materials that are used in a specific insulation system and the chemical interaction of these materials. The determination of these values requires an extensive and complex test procedure. Even then, it requires considerable experience to translate these data to actual motor life. However, for the purposes of this analysis, to estimate the affect of abnormal temperatures on motor life for the life-cycle calculations, it is reasonable to assume that the insulation life (or motor life) decreases by one-half for every 10 to 12 °C increase in operating temperature.

7.3   DIRECT SAVINGS AND PAYBACK ANALYSIS

The annual cost savings for two motors of different efficiencies operating at the same load can be calculated as follows:

$$S = 0.746 \times hp \times P \times H \left(\frac{100}{E_2} - \frac{100}{E_1} \right)$$

where

S = annual saving, \$/yr

hp = horsepower output

P = power costs, \$/kWh

H = running time, hr/yr

E_1, E_2 = efficiencies of motors or systems being compared

Consider the example of a 25-hp motor operating 4000 hr/yr at a power costs of \$0.05/kWh. Calculate the annual savings for an energy-efficient motor with 91.8 nominal efficiency compared to a standard motor with 88.0 nominal efficiency:

$$\text{Annual saving} = 0.746 \times 25 \times 0.05 \times 4000 \left[\frac{100}{88} - \frac{100}{91.8} \right]$$

$$= \$175.46/\text{yr}$$

The years to pay back the premium cost for the higher-efficiency motor can be calculated:

$$\text{Payback in years} = \frac{\text{premium cost}}{\text{annual saving}}$$

Premium cost = cost difference between the two motors or systems being compared for the 25-hp motor in this example,

$$\text{Standard TEFC motor} = \$762$$
$$\text{Energy-efficient TEFC motor} = \$899$$
$$\text{Payback in years} = \frac{899 - 652}{175.46} = 0.78 \text{ yr}$$

The annual savings and the payback period can vary substantially with different power costs and hours per year of running time. For the 25-hp motor example, the annual savings can be calculated for various power rates and operating hours and the payback time calculated. These data can be developed as a series of curves, as shown in Fig. 7.1. Note that with high power costs and relative low operating hours,

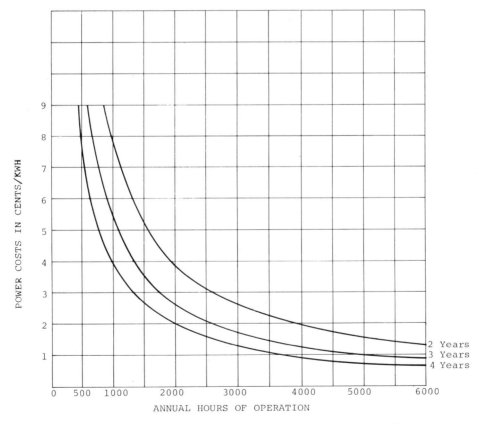

Fig. 7.1 Payback of premium cost for various power costs and operating times for a 25-hp, 1750-rpm motor.

Table 7.3 Unit Energy Saving in Dollars per Horsepower[a]

Lower efficiency	Higher efficiency											
	72	74	76	78	80	82	84	85	86	87	88	89
70	0.296	0.576	0.841	1.093	1.332							
72		0.280	0.545	0.797	1.036	1.264						
74			0.265	0.517	0.756	0.984	1.200					
76				0.252	0.491	0.718	0.935					
78					0.239	0.467	0.683	0.788				
80						0.227	0.444	0.549	0.65	0.750		
82							0.217	0.321	0.423	0.523	0.620	0.716

Lower efficiency	Higher efficiency (continued)										
	85	86	87	88	89	90	91	92	93	94	94.5	95
84	0.704	0.207	0.306	0.404	0.499	0.592	0.683	0.772				
85		0.102	0.202	0.299	0.394	0.488	0.579	0.700	0.755			
86			0.100	0.197	0.292	0.385	0.477	0.566	0.653	0.738		

	91	91.5	92	92.5	93	93.5	94	94.5	95	95.5	96	96.5
87				0.097	0.193	0.286	0.377	0.466	0.553	0.639		
88					0.095	0.188	0.279	0.369	0.456	0.541	0.585	0.625
89						0.093	0.184	0.273	0.361	0.446	0.488	0.529
90							0.091	0.180	0.267	0.353	0.395	0.436

Higher efficiency *(continued)*

	91	91.5	92	92.5	93	93.5	94	94.5	95	95.5	96	96.5
90.5	0.045	0.090	0.134	0.178	0.222	0.264	0.307	0.349				
91.0		0.045	0.089	0.133	0.176	0.219	0.262	0.304	0.345			
91.5			0.044	0.088	0.132	0.174	0.217	0.259	0.300	0.341		
92.0				0.044	0.087	0.130	0.173	0.215	0.256	0.297	0.338	
92.5					0.043	0.086	0.129	0.171	0.212	0.253	0.294	0.334
93.0						0.043	0.085	0.127	0.169	0.210	0.251	0.291
93.5							0.042	0.084	0.126	0.167	0.208	0.248

[a]Based on 1000-hr/yr operation and 1.0¢/kWh power costs.

short payback periods are possible. Also, conversely, with low power costs and relatively high operating hours, short payback periods are also possible.

Table 7.3 has been developed to facilitate the calculation of the annual energy saving between motors of different efficiencies:

$$\text{Dollar annual savings} = \frac{\text{hp} \times 746}{1000} \left[\frac{1}{E_L} - \frac{1}{E_H} \right] \times H \times P$$

where

E_L = lower efficiency
E_H = higher efficiency
H = operation, hr/yr
P = power cost, $/kWh

Table 7.3 derives a unit saving based on 1 hp, 1000 hr of operation per year, and an 0.01 power cost in dollars per kilowatt-hour:

$$\text{Unit savings} = 7.46 \left[\frac{1}{E_L} - \frac{1}{E_H} \right]$$

Then, to compute the dollar annual saving for any horsepower and operating conditions,

$$\text{Dollar annual savings} = \text{unit savings} \times \text{hp} \times \frac{\text{operating hours}}{1000}$$
$$\times \text{power cost in cents}$$

The use of Table 7.3 is best illustrated by the following example.

EXAMPLE

What is the annual savings for a 25-hp motor with 92 percent efficiency over a motor with 88 percent efficiency with 3000 hr of operation per year and power costs of 5¢/kWh?

From Table 7.3 for the efficiencies of 88 to 92 percent the unit saving = 0.369. Then

$$\text{Dollar annual savings} = 0.369 \times 25 \times 3 \times 5 = \$138$$

For the same operating conditions, if the motor efficiencies being compared are 89 to 92 percent, respectively, what is the annual savings?
From Table 7.3, the unit saving = 0.273:

$$\text{Dollar annual savings} = 0.273 \times 25 \times 3 \times 5 = \$102$$

As a guideline to the payback periods to be expected on energy-efficient motors, Table 7.4 has been developed to show the annual sav-

Table 7.4 Annual Dollar Savings and Months to Payback of Investment in Energy-Efficient Motors[a]

hp	Nominal efficiency of standard motors from Table 2.1	Nominal efficiency of energy-efficient motors from Table 2.4	Annual savings (kWh)	Annual savings ($)	Months to payback
1	73	83	492	30	4
2	77	83	560	34	5
3	80	86	781	47	10
5	82	87	1,046	63	9
7.5	84	88	1,211	73	11
10	85	89	1,578	95	10
15	86	90	2,313	139	9
20	87.5	90.5	2,261	136	9
25	88	91.5	3,243	195	8
30	88.5	92	3,848	231	7
40	89.5	92.5	4,325	260	7
50	90	93	5,348	320	6
60	90.5	93	5,318	320	6
75	91	93.5	6,576	394	5
100	91.5	94.0	8,673	521	5
125	92	94.0	8,626	519	6
150	92.5	94.5	10,241	616	7
200	93.0	94.5	10,186	610	9

[a]Based on 4000-hr/yr operation and 6¢/kWh power cost for a dripproof, four-pole, three-phase motor.

ings in kilowatt-hours and in dollars and the payback period in months. This table is based on the average standard motor efficiencies shown in Table 2.1 and average energy-efficient motor efficiencies shown in Table 2.4 for the following conditions:

Standard dripproof construction
4000-hr/yr operation
Cost for electric power: $0.06/kWh

For these operating conditions the payback period for all horsepower ratings is less than 1 yr.

7.4   EFFICIENCY FACTOR OR EFFICIENCY EVALUATION FACTOR

In many industries, such as the process industries, large groups of motors run the same number of hours per year at the same power cost. To facilitate the comparison of various motors under these conditions, an efficiency factor can be developed based on the expected operating life in years (or the life cycle), the cost of power, and the hours per year of running time. This efficiency factor can be expressed in dollars per kilowatt. Thus,

$$EF_{kW} = PNH$$

where
        P = power costs, $/kWh
        N = number of years of operation
        H = hours per year of running time
Or it can be expressed in dollars per horsepower:

$$EF_{hp} = 0.746PNH$$

The factor can be established once for a given group of motors. However, the efficiency factor can vary widely since each of its elements vary over a wide range. Table 7.5 illustrates the range of the horsepower efficiency factor (EF_{hp}) for a 10-yr life cycle at various power costs and hours per year of running time.

When the efficiency factor has been established, the life-cycle power cost savings (LCPS) can then be calculated from the following equation:

$$LCPS = EF_{hp} \times hp \left(\frac{100}{E_1} - \frac{100}{E_2} \right)$$

Table 7.5 Efficiency Factor in Dollars per Horsepower for 10-Yr Life

Power cost, P ($/ kWh)	Annual operation (hr)	Efficiency factor, EF_{hp}	Annual operation (hr)	Efficiency factor, EF_{hp}	Annual operation (hr)	Efficiency factor, EF_{hp}
0.02	1000	149	4000	596	8000	1193
0.03	1000	224	4000	895	8000	1790
0.04	1000	298	4000	1193	8000	2387
0.05	1000	373	4000	1492	8000	2984
0.06	1000	448	4000	1790	8000	3581
0.07	1000	522	4000	2089	8000	4178
0.08	1000	596	4000	2387	8000	4774

This value is the total projected power cost savings based on the life cycle (or evaluation years) and is not the annual savings.

It provides a method of comparing motors of different efficiencies. Consider the example of the 25-hp motor. What is the LCPS comparison of motors with efficiencies of 91, 91.8, and 92.5 percent when compared to a standard motor with an efficiency of 88 percent, assuming a 10-yr life, 4000-hr/yr operation, and 5¢/kWh power costs? From Table 7.5,

$$EF_{hp} = 1492$$

$$LCPS = 1492 \times 25 \left(\frac{100}{88} - \frac{100}{E_2} \right)$$

For the various efficiencies,

Efficiency (%)	Life-cycle power savings ($)
91	1397
91.8	1754
92.5	2062

This method provides a direct power cost savings for different values of efficiency. This life-cycle saving can be compared to the cost of each motor to determine the most economical investment.

7.5   PRESENT VALUE OR PRESENT WORTH METHOD WITH CONSTANT POWER COSTS

The previous payback methods discussed do not consider the cost of money over the payback period. Money invested today at a specific rate of interest will increase in value at some future date. Or conversely, money earned at some future time must be discounted back to the present. Thus the present worth of a future sum of saving can be expressed as

$$\text{Present worth} = \frac{1}{(1 + i)^n} \times \text{annual saving}$$

where

  i = annual rate of interest

  n = year of saving

Thus $100 savings for 5 yr at an interest rate of 10 percent has a present worth of

$$\$100 \times \frac{1}{(1 + 0.10)^1} = \$90.91$$

$$\$100 \times \frac{1}{(1 + 0.10)^2} = \$82.84$$

$$\$100 \times \frac{1}{(1 + 0.10)^3} = \$75.13$$

$$\$100 \times \frac{1}{(1 + 0.10)^4} = \$68.30$$

$$\$100 \times \frac{1}{(1 + 0.10)^5} = \$62.09$$

$$\text{Present worth} = \$379.27$$

—*not* $500. This has the effect of decreasing the net saving or increasing the payback time.

The present worth (PW) calculation can be simplified for a period of years by the equation

$$Present\ worth\ =\ \frac{(1\ +\ i)^n\ -\ 1}{i(1\ +\ i)^n}\ \times\ annual\ savings$$

For the preceding example,

$$PW\ =\ \frac{(1\ +\ 0.10)^5\ -\ 1}{0.10(1\ +\ 0.10)^5}\ \times\ \$100\ =\ 3.791\ \times\ \$100\ =\ \$379.10$$

Tables are available in most accounting manuals for the present worth factor $[(1\ +\ i)^n\ -\ 1]/[i(1\ +\ i)^n]$. For convenience, Table 7.6 is an abbreviated table of the present worth factor.

Consider the 25-hp motor used for sample calculations and the following data: annual saving $S\ =\ \$175.46$, years of life $N\ =\ 10$ yr, and rate of interest $I_R\ =\ 10$ percent. Then

$$PW\ =\ \$175.46\ \times\ \frac{(1\ +\ 0.10)^{10}\ -\ 1}{0.10(1\ +\ 0.10)^{10}}$$

$$=\ \$175.46\ \times\ 6.145\ =\ \$1078.13$$

Compare this to the $1754 calculated by the LCPS comparison method on page 185.

7.6   PRESENT VALUE OR PRESENT WORTH METHOD WITH INCREASING POWER COSTS

The present worth method just discussed was based on a constant cost for electric power. In view of the continuing projected trend of power cost increases, these increases will have a significant impact on the present worth calculations.

Consider the situation where the annual power savings in kilowatt-hours is constant, but there is a percentage increase in the cost of power each year. The annual cost saving will not be constant but will increase in a geometric pattern each year and can be expressed as a series:

$$S(1\ +\ I_p)\ +\ S(1\ +\ I_p)^2\ +\ \cdots\ +\ S(1\ +\ I_p)^n$$

where
S = initial annual power cost savings
N = year
I_p = annual percent increase in power costs
This assumes that the power rate increases by I_p the first year.

Table 7.6 Present Worth Factor[a]

					Interest					
Year	1%	2%	3%	4%	5%	6%	8%	10%	12%	15%
1	0.990	0.980	0.971	0.962	0.952	0.943	0.926	0.909	0.893	0.870
2	1.970	1.942	1.913	1.886	1.859	1.833	1.783	1.736	1.690	1.626
3	2.941	2.884	2.829	2.775	2.723	2.673	2.577	2.487	2.402	2.283
4	3.902	3.808	3.717	3.630	3.546	3.465	3.312	3.170	3.037	2.855
5	4.853	4.713	4.580	4.452	4.329	4.212	3.993	3.791	3.605	3.352
6	5.795	5.601	5.417	5.242	5.076	4.917	4.623	4.355	4.111	3.784
7	6.728	6.472	6.230	6.002	5.786	5.582	5.206	4.868	4.564	4.160
8	7.652	7.325	7.020	6.733	6.463	6.210	5.747	5.335	4.968	4.487
9	8.566	8.162	7.786	7.435	7.108	6.802	6.247	5.759	5.328	4.772
10	9.471	8.983	8.530	8.111	7.722	7.360	6.710	6.145	5.650	5.019
11	10.368	9.787	9.253	8.760	8.306	7.887	7.139	6.495	5.938	5.234
12	11.255	10.575	9.954	9.385	8.863	8.384	7.536	6.814	6.194	5.421
13	12.134	11.348	10.635	9.986	9.394	8.853	7.904	7.103	6.424	5.583
14	13.004	12.106	11.296	10.563	9.899	9.295	8.244	7.367	6.628	5.724
15	13.865	12.849	11.938	11.118	10.380	9.712	8.559	7.606	6.811	5.847

[a]$PWF = [1 + i)^n - 1]/[i(1 + i)^n]$.

The present worth of this series of savings including the cost of money or required rate of return is

$$PW = S \left[\frac{1 + I_p}{1 + I_R} + \frac{(1 + I_p)^2}{(1 + I_R)^2} + \cdots + \frac{(1 + I_p)^n}{(I + I_R)^n} \right]$$

where I_R is the annual interest rate for money or required rate of return.

An effective interest rate can be determined for the preceding expression which is a function of the percent increase in power costs and the rate of return:

$$I_E = \frac{1 + I_R}{1 + I_p} - 1$$

where

  I_E = effective interest rate
  I_R = annual interest rate or rate of return
  I_p = annual percent increase in power cost
The present worth can then be calculated as follows:

$$PW = S \left[\frac{(1 + I_E)^n - 1}{I_E(1 + I_E)^n} \right]$$

The values of the bracketed expression for the equivalent interest is the equivalent present worth and can be calculated or selected from Table 7.6.

Consider the 25-hp motor. The annual power cost saving = $175.46. For a life of 10 yr with the cost of money at 10 percent and the annual increase in power cost rates at 8 percent.

$$I_E = \frac{1 + 0.10}{1 + 0.08} - 1 = 0.0185 \text{ (or 1.85 equivalent interest rate)}$$

$$PW = \$175.46 \left[\frac{(1 + 0.0185)^{10} - 1}{0.0185(1 + 0.0185)^{10}} \right]$$

$$= \$175.46 \times 9.05$$

$$= \$1587.91$$

Compare this to the $1078.13 calculated for the present worth at constant power costs on page 187.

7.7   NET PRESENT WORTH METHOD

The previous methods do not consider the impact of taxes or depreciation on the present worth in order to determine the *net* present worth of the power cost saving.

The net present worth (NPW) or the justified premium for the energy savings consists of two components:

$$NPW = NPW_E + NPW_D$$

where

NPW_E = net present worth of energy saving
NPW_D = net present worth of the depreciation on the premium investment

and

$$NPW_E = PW(1 - T)$$

where

PW = present worth
T = tax rate

Then

$$PW = S \left[\frac{(1 + I_E)^n - 1}{I_E(1 + I_E)^n} \right]$$

$$I_E = \frac{1 + I_R}{1 + I_p} - 1$$

where S is the annual savings based on the initial power rate.

After the value of the effective interest I_E has been calculated, the value of the bracketed expression, which is the present worth, can be calculated or selected from Table 7.6. The net present worth of the depreciation (NPW_D) can be calculated as follows:

$$NPW_D = \left(\frac{NPW_E}{N} \right) \left[\frac{(1 + I_R)^n - 1}{I_R(1 + I_R)^n T} \right]$$

This assumes straight-line depreciation of the premium cost over the life cycle of the motor.

The PW factor in this part of the calculation is based on the interest cost of money I_R and can be determined from Table 7.6.

Consider the 25-hp motor example and the following data: the life N = 10 yr, the cost of money I_R = 10 percent, the annual power rate increase I_p = 8 percent, the initial annual power saving S = $175.46, and the tax rate T = 40 percent. Then

$$I_E = \frac{1 + 0.10}{1 + 0.08} - 1 = 0.0185$$

$$\text{PW factor} = \frac{(1 + 0.0185)^{10} - 1}{0.0185(1 + 0.0185)^{10}} = 9.05$$

$$\text{PW} = \$175.46 \times 9.05 = \$1587.91$$

$$\text{NPW}_E = \$1587.91(1 - 0.40) = \$952.75$$

$$\text{NPW}_D = \frac{\$952.75}{10} \times 6.145 \times 0.40 = \$234.19$$

where 6.145 is the present worth factor for 10 percent interest and 10 yr. Then

$$\text{NPW} = \$952.75 + \$234.19 = \$1186.94$$

This is the total net present worth of the power saving, or the maximum premium that can be expended to achieve the energy savings.

What is the impact of motor efficiency on this calculation? Consider the example of the 25-hp motor with various efficiency motors available:

Motor	Efficiency (%)	Initial annual saving ($)	NPW_E	NPW_D	NPW
Standard	88				
Premium A	91	139.74	758.79	186.51	945.30
Premium B	91.8	175.46	952.75	234.19	1186.94
Premium C	92.5	206.20	1119.67	275.21	1394.88

The NPW indicates the net present worth of the energy saving at each efficiency level, or conversely indicates the premium motor price that can be justified to achieve the power saving and motor efficiency level.

In a given facility many of the factors that make up the preceding calculation can be considered constant. When this assumption is made, the calculation for comparing the net present worth of motors of different efficiency can be simplified by calculating the constant factors derived as follows: elements assumed to be constant:

Initial power cost ($/kWh), P
Hours of operation per year, H
Equipment life (yr), N

Annual power cost increase, I_p
Cost of money, I_R
Equivalent interest rate, I_E
Tax rate, T
Rated motor horsepower, hp

Let

$$K_1 = 0.746PH \left[\frac{(1 + I_E)^n - 1}{I_E(1 + I_E)^n} \right] (1 - T)$$

where

$$I_E = \frac{1 + I_R}{1 + I_p} - 1$$

and

$$K_2 = K_1 \frac{1}{N} T \left[\frac{(1 + I_R)^n - 1}{I_R(1 + I_R)^n} \right]$$

Then

$$NPW_E = hp \left(\frac{100}{E_1} - \frac{100}{E_2} \right) K_1$$

$$NPW_D = hp \left(\frac{100}{E_1} - \frac{100}{E_2} \right) K_2$$

The net present worth then simplifies to the following:

$$NPW = NPW_E + NPW_D$$
$$= hp \left(\frac{100}{E_1} - \frac{100}{E_2} \right) (K_1 + K_2)$$

Again consider the 25-hp motor of previous examples:

$$K_1 = 0.746 \times 0.05 \times 4000 \left[\frac{(1 + 0.0185)^{10} - 1}{0.0185(1 + 0.0185)^{10}} \right]$$
$$\times (1 - 0.40) = 810.16$$

$$I_E = \frac{1 + 0.10}{1 + 0.08} - 1 = 0.0185$$

$$K_2 = 810.16 \times \frac{1}{10} \times 0.40 \left[\frac{(1 + 0.10)^{10} - 1}{0.10(1 + 0.10)^{10}} \right] = 199.14$$

$$NPW = 25 \left(\frac{100}{88} - \frac{100}{91.8}\right) (810.16 + 199.14) = 1186.92$$

Again this expresses the total net worth of the power saving, or the maximum premium that can be expended to achieve the energy saving.

Other methods can be and have been used to evaluate the energy economics when comparing motor or system efficiencies and power savings. However, considering the long-range assumptions that have to be made for any of the methods, the net present worth method, which considers the major factors that contribute to such a calculation, is considered a sound and conservative procedure for evaluating energy economics in industrial systems.

Selected Readings

Alger, P. L., *The Nature of Induction Machines*, Gordon & Breach, New York, 1965.

ASHRAE, *Guide and Data Book—Equipment, 1969*, ASHRAE, New York, 1969.

Bedford, B. D., and R. G. Hoft, *Principles of Inverter Circuits*, Wiley, New York, 1964.

Bose, B. K., *Adjustable Speed AC Drive Systems*, IEEE Press, New York, 1981.

Control Engineering, *Energy Conservation*, Dun-Donnelley, Chicago, 1977.

Harnden, J. D., Jr. and F. B. Golden, *Power Semiconductor Applications*, Vol. I, General Considerations, IEEE Press, New York, 1972.

Hicks, T. G., and T. W. Edwards, *Pump Application Engineering*, McGraw-Hill, New York, 1971.

Jaeschke, R. L., *Controlling Power Transmission Systems*, Penton/IPC, Cleveland, 1978.

Karassik, I., and R. Carter, *Centrifugal Pumps*, McGraw-Hill, New York, 1960.

Karassik, I. J., W. C. Krutzsch, W. H. Fraser, and J. P. Messina, *Pump Handbook*, McGraw-Hill, New York, 1976.

Lloyd, T. C., *Electric Motors and Their Applications*, Wiley-Interscience, New York, 1969.

NEMA Standards Publication MG10-1977 *Energy Management Guide for Selection and Use of Polyphase Motors*, National Electrical Manufacturers Association, Washington, D.C., 1977.

NEMA Standards Publication MG11-1977, *Energy Management Guide for Selection and Use of Single Phase Motors*, National Electrical Manufacturers Association, Washington, D.C., 1977.

NEMA Standards Publication MG1-1978, *Motors and Generators*, National Electrical Manufacturers Association, Washington, D.C., 1978.

Rachlin, R., *Return on Investment, Strategies for Profit*, Prentice-Hall, Englewood Cliffs, N.J., 1979.

Ramshaw, R. S., *Power Electronics, Thyristor Controlled Power for Electric Motors*, Chapman & Hall, London, 1973.

Say, M. G., *Alternating Current Machines*, Halsted Press, New York, 1976.

Shupe, D. S., *What Every Engineer Should Know About Economic Decision Analysis*, Dekker, New York, 1980.

Smeaton, R. W., *Motor Application and Maintenance Handbook*, McGraw-Hill, New York, 1969.

Smith, G. W., *Engineering Economy, Analysis of Capital Expenditures*, Iowa State University Press, Ames, 1979.

U.S. Department of Commerce, ERDA-76/130 Rev., *Life Cycle Costing, Emphasizing Energy Conservation*, Washington, D.C., 1976.

U.S. Department of Energy, DOE/EIA-0040(79), *Typical Electric Bills January 1, 1979*, Washington, D.C., Oct. 1979.

U.S. Department of Energy, DOE/EIA-0173(79)/3, Vol. 3, *Annual Report to Congress*, Washington, D.C., 1979.

U.S. Department of Energy, DOE/TIC-11339, *Classification and Evaluation of Electric Motors and Pumps*, Washington, D.C., 1980.

Veinott, C. G., *Theory and Design of Small Induction Motors*, McGraw-Hill, New York, 1959.

Weston, J. F., and E. F. Brigham, *Managerial Finance*, Dryden Press, Hinsdale, Ill., 1975.

Index